NO TWO DAYS THE SAME

BRINGING BACK A DIVISION

MY TIME WITH THE 29TH ID(L) 1985 — 1990

JOE E. HARRIS, JR.

FOREWORD BY WILLIAM S. MUND, JR.

No Two Days the Same: Bringing Back a Division

29th Infantry Division (Light) 1985-1990

A Memoir by Joe E. Harris, Jr.

Published by JWC Publishing

JWC PUBLISHING
WHAT'S *YOUR* STORY?

ISBN: 978-1-951797-88-1

Printed in the United States of America

Dedication

Colonel Edward "Craig" Norman
1947-2022

Assistant Chief of Staff, G3 29th Infantry Division (Light) 1985

Table of Contents

Foreword

It was a cold winter's day in Northern Virginia on Friday, 4 January 1985. Snow had fallen during the night and Ft. Belvoir was a sheet of white snow and ice. Luckily for us, at North Post, we still had heat, electricity, and running water, as this was always a constant concern for us working in Building 2247. This dilapidated WWII building was at this time, the provisional headquarters of the 29th Infantry Division (Light); a unit that would be officially reactivated later on in the year.

Some of us had been assigned to this headquarters in mid-November 1984 and had been working in this building and this North Post area under austere conditions. Personally, I had the distinction of being the first enlisted AGR soldier assigned to this headquarters.

It was on this cold winter's day that Captain Joe E. Harris, a native of North Carolina and currently residing in the Tidewater area of Virginia, first set foot on Fort Belvoir. He arrived about noon and I was the first person he met. He informed me that he was the latest addition to the staff of the provisional division headquarters.

I welcomed CPT Harris enthusiastically not knowing anything about him or where he was from, etc. As I was the Executive Administrative Assistant serving directly under the Chief of Staff, Colonel William C. Bilo, it was my duty to escort new arrivals down the hall to meet the CofS in his office. I had briefly served under COL Bilo when he was

Director of Personnel, MDARNG, at the Fifth Regiment Armory in Baltimore, MD.

My first impression of CPT Harris was that he seemed to be about the same age as me. He displayed and spoke with all the attributes of a proper Army officer. In speaking with him later, I realized that he didn't really know anyone and he was going to be away from his wife, family, and home for an extended period of time. In spite of these difficulties, he seemed to be eager and ready for the myriad of challenges which he and the rest of us would encounter with the actual formation and reactivation of a reserve component light infantry division.

CPT Harris and I got to know each other better as the winter wore on and eventually turned into spring. It was sometime during this early spring period that CPT Harris received federal recognition of his promotion to Major.

During some of this time both he and I had adjoining rooms in a BOQ building that we were allowed to stay in. This meant that we both shared a toilet and shower stall that was located between the two rooms. I kind of liked this setup because in addition to the bedroom which contained a sink for washing and shaving, it also had a smaller room with a table and full-size refrigerator. It was a sad day indeed when we were evicted to make room for the upcoming annual training period. This building would be inhabited by Reserve or National Guard army officers who were doing their annual training at Fort Belvoir.

After I left the Division Headquarters, we both sporadically kept in touch. With the advent of the Internet, email, etc., it became much easier for us to communicate. I was quite honored and surprised when LTC (Ret) Joe Harris asked me to write the foreword to his book. When I asked him why he thought I should write it, his answer was, "because you were there."

Yes, I was there and I will end on that note, and graciously thank him for giving me the opportunity to write something that I never thought someone would ever ask me to write.

WILLIAM S. MUND, JR.
SSG (Ret), USAR
B/1/115 & HHC/29ID(L)
MDARNG/VAARNG

Prologue

As the United States prepared its military for entrance into World War 1, the 29th Division was constituted on paper in the summer of 1917. The division's first headquarters was located at Camp McClellan, Alabama. Following nine months of training in Alabama, the division prepared for deployment. In late July of 1918, the division moved to the Western Front where it was engaged in several offensives. The division played a key role in the Meuse-Argonne offensive and fought around such key terrain features as Malbrouck Hill, Molleville Farm and the Grand Montaigne and Etrayes Woods. The Armistice ended the war on November 11, 1918. When it was finally over, the 29th Division had suffered 5,552 casualties. Following the end of the war, the division reverted to National Guard control.

The 29th Division was federalized and mobilized for World War II on February 3, 1941. The division was stationed at Fort Meade, Maryland where it trained for deployment. While there, the division reorganized under the "triangular" structure and changed its designation to the 29th Infantry Division. In the early fall of 1942, it departed by ship for Scotland. The division then moved to southern England and began preparing for an amphibious attack at Normandy, France. The now famous D-Day invasion on the Normandy beaches began on the morning of June 6, 1944. The 29th Infantry Division landed and attacked on what is now known as bloody Omaha Beach. Over the next several months, the division fought its way through the hedgerow country of Normandy. By late summer the division was sent to Brittany. The enemy stronghold at

Brest soon surrendered and the division was moved again to assist in the fighting around Aachen, Julich and Koslar. Following V-E Day, units of the division began returning home to the US. The 29th Infantry Division was inactivated at Camp Kilmer, New Jersey on January 17, 1946. Between then and now the 29th Infantry Division, a division in the National Guard, underwent several reorganizations. Changing requirements in the nation's ground combat forces eliminated the need for the division, and in 1968, it was eliminated all together.

On June 6, 1984, the Secretary of Defense announced the decision to reactivate the 29th Infantry Division. This time the division would be designated the 29th Infantry Division (Light). Division headquarters would be stationed on Fort Belvoir, Virginia.

I reported to the temporary division headquarters on January 4, 1985. All the episodes in this book occurred. They are a collection of my experiences and remembrances while working and living on Fort Belvoir. Perceptions and opinions are the author's. They do not represent the attitude of the 29th Infantry Division (Light).

Introduction

Reading the Army Times in the summer of 1984, I discovered an article on the front page that drew my interest. The announcement by the Secretary of Defense, Casper Weinberger, of the planned reactivation of the 29th Infantry Division took place in Washington, D.C. on the 40th anniversary of the D-Day invasion, June 6, 1984. It was to be organized as a light infantry division.

The Cold War had been on for years. The Army's war fighting doctrine, the Air Land Battle, was centered mostly around the use of armor (tanks) and mechanized infantry divisions in the European Theater. A new Chief of Staff of the Army (CSA) was appointed, and change was afoot. He declared that four "heavy" divisions would turn in their armored personnel carriers and become light or foot-mounted infantry. A fifth light division would come from the resources of the National Guard.

There was heavy bidding from the National Guard for this one division, but in the end, the Secretary of the Army (SOA) selected the "Blue and Gray" — the 29th Infantry Division of D-Day fame. The Maryland and Virginia National Guards would join forces again to form the 29th Infantry Division (Light).

My command of a mechanized rifle company was demanding. Exhausted from stress and general fatigue of command, I had been delighted by the reassignment to the US Army Cadet Command. Detailed to the Department of Military Science at Old Dominion University in Norfolk,

Virginia, I was an instructor of introductory classes for freshmen enrolled in Army ROTC. It was not a challenging nor was it a particularly interesting job, but it did afford me time to work on a master's degree at night school. Now, a full three years into this mostly paper pushing job in Norfolk, Virginia, I was ready to shift into high gear again. I sensed opportunity with the reactivation news of the 29th Division.

I called the office of The Adjutant General (TAG) and requested a personal interview. TAG—Virginia referred my request to his Deputy Chief of Staff for Personnel. I was granted an appointment to discuss job opportunities on the new division staff.

At the interview I was told that, yes, there were positions to be filled on the 29th Division's general staff. The Assistant G3 (operations & training) and Assistant G4 (logistics) were already identified, but candidates were sought for the Assistant G2 (intelligence & security) and the Assistant G1 (personnel & administration). Both positions were at the pay-grade of 04, meaning a major's billet. It meant a possible promotion could be in the cards. I left the state headquarters feeling confident that I had put my best foot forward. I was encouraged to apply for both positions. I did.

I had a strong infantry background for a company grade officer and thought my chance at the Assistant G1 job was reasonably good. I was not qualified as an intelligence officer and gave this position little chance. But still, if offered, I was willing to be trained in the field of tactical intelligence. It would be a big leap and a branch transfer after years of underpinning as an infantry officer. I prayed for the Assistant G1 position, not remembering to be careful of what you pray for.

A couple of weeks later I received a letter from the personnel section. Both positions had been filled by others. I was thanked for applying, but no cigar. Seething from disappointment, I decided to wait a few days before calling state headquarters to express my disappointment over not being given an opportunity to at least be interviewed. I knew this new division would shake up the organization and thus create other positions at brigade and battalion levels. I wanted to keep my foot in the door for a future position in the new light division.

A few days passed. I received a telephone call from someone of whom I had never heard. With that phone call, my military trajectory was about to change. Colonel William C. Bilo asked if I still had an interest in a position on the division staff. I excitedly answered that I did. He explained that the Assistant G2 position was available. "I have a letter saying it is already filled," I said. He answered that the situation had changed and that he, division chief of staff, was authorized to fill the billet. He went on to say that he liked what he saw in my officer dossier.

I explained that I wasn't branch qualified for the assignment. Colonel Bilo then said it would take at least a full year before I had the security clearance for schooling. In the meantime, he needed someone at that level who could be useful in a myriad of challenging duties. I told him I was looking for just that sort of learning experience. "How soon can you get here," he asked? "Two weeks," I said.

It snowed in northern Virginia the night before I departed Tidewater for Fort Belvoir. On January 4, 1985 I skidded into the iced-over installation shortly after 1200. In uniform I fussed to myself about messing up my highly polished boots as I walked from the loaded car to the entrance of Building #2247, a two-story WWII barracks.

Entering the temporary division headquarters (there was no signage denoting division headquarters) I stomped the slush from by boots thereby making my presence known. I was greeted by SGT William S. Mund, Jr.

SGT Mund was pleased to see me. He said, "We've been eagerly awaiting your arrival, sir. We thought you would report first thing this morning." As we spoke I could hear shouting down the narrow hall at the opposite end of the building. It was distractingly loud. SGT Mund pointed with his head in the direction of the commotion. "That's the chief's office down there. He's engaged in conversation at the moment with Lieutenant Colonel Norman, the G3," he said. Soon the shouting stopped, and a door slammed shut.

Staring at me was Lieutenant Colonel Edward C. Norman. Tall with premature white hair, his face stood out as red as a beet. Glancing down at my name tag, he reached out to shake my hand. "Welcome aboard Harris, if you need anything I'm in the next building across the street from this one. Good luck," he said as he quickly exited the building.

Fear was setting in as I figured I was next for a good butt chewing. Colonel Bilo's mood was already established by the previous officer's visit, and I had been expected earlier. What had I gotten myself into? I didn't have long to dwell on the unknowns as SGT Mund escorted me to the chief's office. "Colonel Bilo," he announced, "meet our final member of the staff, Captain Harris." The introduction was nothing like I anticipated. The chief's demeanor was calm, cool, and collected. He asked me to sit down. I complied, and we talked about many issues. He began by saying, "You'll have a great boss. Lieutenant Colonel Wayne D. Zajac is the G2. A smart and very clever officer, you shouldn't have any problems. He'll be a great mentor."

The chief gave me a quick description of my initial duties. "Since it'll take the better part of a year to clear you for training at Fort Huachuca, you will set up and operate the office of G6." He could see my puzzled expression so he elaborated. "The G6 is also Lieutenant Colonel Zajac, and you're his assistant. For the first year or so there will not be an intelligence mission here, so your job is to manage the office of transition." He continued, "This office will set the schedule for unit conversions in all matters from their present configuration to their newly reorganized one in the light division's configuration." He stood up and said, "Come with me."

Exiting through a side door we crossed the street and entered another old dilapidated WWII building. G2 and G6 were on the first floor. The G3, Lieutenant Colonel Norman, and team already occupied the second floor. As we walked around the chopped up bay (converted into individual rooms years ago), the chief opened a door and said, "This is it, you'll establish the office of transition right in here."

I looked around the room sizing up its dismal condition. The chief said, "Doesn't look like much now, but refurbished government desks, tables, and chairs are on order and will be here any day, so will the telephones. In the meantime, do the best you can. Joe, do you have a typewriter and a cot? Good! I'll locate a chair for you and table for the typewriter. We gather at 1600 in my office to discuss daily progress and plans for tomorrow." With that, the chief stepped out the room, walked down the hall and hearing the door open and shut, I assumed he departed.

This was the meager beginnings of the 29th Infantry Division (Light). The office of transition contained one bare 60-watt lightbulb in the ceiling. When I pulled the string, it broke away from the fixture. There was a metal chair propped against the wall. One leg was bent out of shape and when opened it simply fell over. Also visible was a broken 2X4 laying on the floor and a cinder block set in the corner.

I tried to make sense of it all. It seemed like a sick joke. But then I reminded myself that we were gathered here for a huge project — the restructuring of nearly all of the army assets of two states into a light infantry division for reactivation. The one bright spot about Building #2246 was that the oil fueled furnace worked and provided plenty of hot water and heat.

This place would be my office and living quarters until other more appropriate arrangements could be made. This was ground zero. There was only one direction to go, up. Seeing opportunity, I had asked for a challenge. One had been delivered — coordinate the massive conversion of equipment and personnel into a new 10,600-person light infantry division. I didn't know a soul, but I was onboard in a major's position as an unqualified Assistant G2 now in charge of an office for transition (G6), a function I had never heard of in all my military education or training.

During the five years I was posted at the division headquarters, I had several assignments. Those included Assistant G1, Secretary of the General Staff, Assistant G3 and finally as a trained and qualified military intelligence officer in the G2. There's a good reason an infantryman is often called a jack of all trades. I was about to show how.

CHAPTER ONE

G6

It was called the office of transition and labeled G6. This was not a notional concept. It was the real deal. Staff sections are usually broken down by their functions. Personnel in a tactical command is known as S1. Intelligence/security is S2. Operations/training is S3. Logistics is S4, and Civil Affairs (sometimes referred to as Civil Military Operations) is S5. However, at the division level, the letter G represents general, for the general staff. The number six (6) is assigned to a temporary function. The function of transition was always meant to be a limited period. Then it would be dissolved. The G6 was an ad hoc, temporary function. It would last as long as needed to oversee the conversion of units to the light division's table of organization and equipment (TO&E).

The G6 acted as the umbrella for the entire conversion process. Its responsibilities included the monitoring and coordination of all activities associated with the development of the new light division.

Colonel Zajac and I took most of our direction and all of our transition guidance from the chief of staff. However, the principal staff, G1, 2, 3, 4, and 5, played major roles. Interfacing with those staff sections was paramount. We got to know one another quickly as the tasks involved for reactivation turned out to be more operose than thought. Not to be too hard on ourselves, none of us had done anything like this before. As for me, each day was filled with something new. It was an exciting time, and I felt at the center of it all.

WHITEBOARD

The staff met regularly to brief-back the day's events, however, one day we met for a reactivation brainstorming session. It was time for some reverse planning. (Reverse planning is a technique first used by the aviators. Essentially, the plan begins at the end and moves through the steps in reverse order up to the present.) When I arrived for the meeting, the chief told me to use the whiteboard to record the salient points, but there was an issue. I was left-handed.

From my childhood I hated being sent to the blackboard. I couldn't write with chalk without smearing it with the back of my hand. I was messy, and my classmates knew it. Even my papers written in pencil were smeared from soft lead. As the meeting began, I recorded what I considered to be relevant information. As a lefty, my body blocked the view of the board as I swiftly wrote with a magic-marker trying hard not to smear the work. It was barely legible, but I was doing the best I could. Without warning, the chief abruptly leaped from his seat, snatched the marker from my hand and pushed me aside. I might have fallen had it not been for a wall that stopped my off-balance tumble. My contemporaries stared at me in disbelief as the chief erased the board. I remember one of them was slowly shaking his head and making eye contact at me as if to say, be cool. I took my seat among them as the chief became the scribe. I felt humiliated, but I never said anything about my feelings. It was a rough morning, but I soon moved on rationalizing that they were relieved that it didn't happen to them.

Nothing was ever said or done about my public display of weakness as a note taker, and for the rest of my military career, I never used a white board again.

TECHNIQUE

The winters can be very cold outside of Washington, D. C. With the Potomac River nearby, the air was always humid in any season. I initially lived in my office, and rested at night on a cot brought from home. Though the cot was adequate, any quality sleep was from my own bed at home in Tidewater Virginia. Fatigue became a factor. Eventually better housing was acquired about a quarter mile from the headquarters. I was now sleeping on a mattress with a sink in the room. There was a shower and commode between adjoining rooms. The building next door had a washer and dryer. Conditions were improving ... then we lost power during a heavy ice storm. I remember the rude awaking. I could see my breathe in the mirror as I quickly dressed. A shower and shave was out of the question.

I warmed up in my car, but the headquarters area was without power too. This meant that my electric typewriter was useless. All progress on reactivation matters came to a screeching halt, so I thought. The walk from my office to headquarters was a precarious one. On entering the building, I heard a typewriter. To my astonishment there was SGT Mund feverishly working on the chief's correspondence using a manual typewriter. He was wearing a toboggan on his head and black leather shells (gloves) on his hands.

"The chief isn't here, sir," said Mund. "He left for home late in the afternoon anticipating the weather forecast. There's no electricity and it's getting dark in here."

We were momentarily crippled and stranded. We would just have to wait on the power company to restore electricity. We struggled on. The welcomed news was that the toilets still worked as did the headquarters' telephone.

That evening following an early supper, I bedded down in a heatless room for a long winter's night. Then there was a knock on the door. SGT Mund occupied the room opposite the shower so I surmised it was him.

"Sir," said Mund.

"The chief said to run water in your sink and use the drain stopper to hold it there overnight."

"What good will that do?" I answered.

Mund responded, "When you shave in the morning the water will be at least room temperature. The water in pipes might freeze overnight."

UTAH

The winter of 1985 was a typical frigid one. Plunging temperatures and dampness from freezing rain, ice, sleet, and snow ensured that traffic around the beltways were adventures. The Assistant G1 and I were on our way to Baltimore, Maryland early one morning for a meeting on personnel transition matters at the state headquarters. Neither of us were familiar with the great city. The chief, a

Marylander, gave us some instructions before we took off. I distinctly remember him saying to us, "Be sure to take the Utah exit from the freeway." (Note, homophones can lead one astray.) We reached the city limits without incident amid constant snow flurries between Washington and Baltimore.

I never saw the exit for Utah Street. By the time we crossed the Patapsco River, I realized we would not make the meeting on time. It took another half hour to turn around in Dundalk. Crossing the river again was a cinch, but this time we exited at the first ramp leading into the city. I remember saying to the Assistant G1 that infantry officers are never lost, only momentarily disorientated. "We are lost," he replied.

It began snowing. Judging by the size of the flakes smashing into the windshield, I guessed the temperature was just above freezing. The Fifth Regiment Armory was our inner-city destination.

Finding the armory, I was awed by its immense size and design. It appeared to me like a stone fortress with bastions at each corner. Stepping through its massive twin iron door entrance, we were registered as visitors and given a security pass with directions to the second floor. I heard a whistle and cheers from people inside.

I was amazed to discover an indoor track and field meet in full progress under roof. There were sections of the floor set aside for jumping and throwing events, and, of course, runners racing in various distance events around the perimeter of the armory floor. An indoor track and field meet in the dead of winter. This was a first for me.

We missed the meeting and rightly so but made up for it before we drove away. On our way out of the city we spotted the street sign for Eutaw. Who knew(?) — everybody from Maryland, I guess? The joke was on us and we laughed all the way home to Fort Belvoir, Virginia.

#268

One morning we began our day with no running water. Somewhere on post a water main had burst and unfortunately, the latrine was already soiled by the time someone affixed an out-of-order sign on the door. As time passed, foul odors contaminated the headquarters. To make matters worse, the post engineer considered these WWII barracks buildings low priority. At some point the chief loaded us up in the staff car, and we motored over to Fort Belvoir's officer club to use the latrine. An infantry division headquarters on a federal post didn't receive any respect from the post commander.

Days passed without running water. The air inside of division headquarters became foul. The absence of running water continued. Then the US First Army commander called to set up a visit at the division headquarters. As an army commander, he was responsible to the Chief of Staff of the Army (CSA) for the training and development of units in his area of interest. (The idea of light divisions was the brainchild of this CSA.) He was flown in his airplane from Fort Meade over to Fort Belvoir. I drove the staff car for the chief down to the Davidson Army Airfield. Colonel Bilo met the general on the tarmac and escorted him to the staff car and away we drove back to division headquarters on Fort Belvoir. While we traveled, the chief attempted to address his concerns for lack of post support. We shortly arrived at our WWII buildings.

The US First Army commander met the staff and was briefed by them. Then he disappeared with the chief into an office where they spent the better part of an hour behind closed doors. My job was to stick around to transport the general back to the airfield. When the two officers came out, plans had changed. Rather than returning to the airfield, the chief told me to take them to the post commander's headquarters.

I parked the car next to the post commander's personal vehicle. The chief and US First Army commander walked down the long sidewalk leading to the main entrance to post headquarters. (At that time, Fort Belvoir was the Engineer Center for the army and was under the command of a major general — an engineer himself.)

I waited patiently in the car for my passengers to return. When they appeared on the front steps, I started the ignition. Once inside, the chief told me to take them straight to the airfield. The trip was short, and not much was said. I could tell that the general's mind was preoccupied.

On the return trip to division headquarters, the chief shared a little of what had occurred at the post commander's office visit. He told me that the US First Army commander had told him to wait in the building's vestibule. His visit with the post commander would be a private conversation, general to general. That was about it, but the chief did smile and say that some relief to our lack of support was about to happen. As we arrived at Building #2247, much to my surprise, a plumber's truck was parked nearby. We had flush toilets before the end of business in our old WWII two- story building.

Not long after the US First Army commander's visit to Fort Belvoir, we were briefed on some exciting developments concerning the division headquarters. The US First Army commander had definitely "pulled rank" on his subordinate post commander for demonstrating little interest in the reactivation of the 29th Infantry Division (Light). It was my understanding that the Division Council (both TAGs — Maryland and Virginia and three other generals) would meet at division headquarters on a specified date to select a suitable site for the new division headquarters.(1) The only building off limits to the council was the post headquarters building. The council studied several sites, but finally settled upon Building 268 (Williams Hall) as its location was next to the post commander's and across the street from Long Parade Field.

As it turned out, even Williams Hall was too small to support a division headquarters and headquarters company. But the council had done its job and the post commander, well, he had to look at his debacle everyday going to and leaving work. In fact, the division headquarters building now received attention and respect that had been long overdue. The National Guard Bureau allocated about a quarter of a million dollars for the renovation of Williams Hall.

MAMMOTH

The chief and I were in Richmond, Virginia attending transition meetings at the state headquarters. Transition was not new to the Virginians. Units came and went according to decisions made by the "higher ups." Armories were always needing renovation or expansion and in isolated cases sold away when no longer needed to house a unit. The development of a new brigade certainly challenged the staff to work out solutions. 2nd Brigade would be comprised of two infantry battalions yet to be organized in Virginia. A third infantry battalion had already been identified from Maryland. The meetings were grueling for these were the days when smoking was allowed in government facilities. I detested the air in the conference rooms. Jelly filled donuts and strong coffee reigned supreme along with ashtrays and powdered milk substitute. Forget bottled water, juice, and fruit assortments. This was back in the day when men were men.

At the conclusion of the smoke drill, the colonel and I rode in a staff car to a motel where we checked in for the evening. I immediately changed into running clothes and tried to get in a few miles before dinner. I coughed and hacked out the mucus caused by second hand smoke for the first mile or so. After a shower and a change of clothes, we took off for dinner.

The chief said that we were going to dine at Satterwhite's Restaurant. That was okay with me. I was ready for something other than a chain and hoped that there was a non-smoking section in the restaurant. Satterwhite's was way out of town. We must have driven over twenty miles maybe thirty before stopping. In fact, we drove out of Henrico and into Goochland County bordering the James River. I wondered why so far? I rationalized that Satterwhite's must be a very good restaurant.

Before pulling into the parking lot, I had already seen the huge sign advertising the world's largest hamburger — The Mammoth Burger! How big was it I pondered?

A waitress seated us and handed me a menu. The chief said, "Bet you can't eat all of The Mammoth Burger." He egged me on and claimed he would pay for it if I ate it all. I looked around the restaurant and saw photographs of those claiming to have eaten a Mammoth Burger. But, how big could it be? Curiosity was killing me. I ordered The Mammoth Burger and waited in anticipation.

When it was served, I was aghast by the size of the burger. I never mentioned melting cheese on it. It included a proportionate amount of lettuce, tomato, and onion piled between a bun the size of a dinner plate. Holding it was hopeless. I resorted to a knife and fork and began working. I worked and worked and worked some more on the hamburger. At the halfway point I was full. How was this other half going to be eaten? I briefly thought about the railing I would receive if I failed, so I began to work again. Little by little The Mammoth Burger began to disappear.

When it was gone a cheer was heard around the room and a waitress took my picture. The chief, true to his word, paid for my dinner. At this point I felt alright. I was not driving so I sat still in the front seat. That's when it got interesting. We had driven out into the heart of the next county, and on the return route I began to feel nauseated. It was dark, and even on-coming headlights bothered me. Potholes and curves were torturous. Then the trip mercifully ended.

Breathing fresh air was exhilarating. I walked for a while to let my digestive system come out of overload. I was satisfied over my most recent accomplishment. I had eaten the whole dang Mammoth Burger thing. That night I slept like a baby.

A few days later back at Fort Belvoir, the G1 asked me how I liked that Mammoth Burger. "How did you know?" I asked. He grinned and then explained that it was a kind of rite of passage. The same prank was played on the chief by members of the state headquarters in Richmond. In hindsight, I never saw the chief's picture on the wall at the restaurant.

ACHIEVEMENT

As winter changed into warmer spring-like weather, I decided it was time to get out of the G-6 for a day and watch some field training near Fort Meade, Maryland. Companies from the 175th Infantry Regiment were engaged in "lanes" training. (This regiment would eventually become the core of the light division's 3rd Brigade.) For the moment, the unit was carrying out its normal training schedule under the separate brigade's training plan. Lanes or terrain had been established to help teach infantry squads negotiate obstacles such as how to breach enemy lines, how to neutralize an enemy bunker, how to attack a trench using fire and maneuvering A and B Teams. This had been my element until joining the division staff. I watched eagerly as the learning process was carried out. I was super charged by the training. It was all I could do to keep from participating. I was just an observer; I had no other role but to keep out of the trainer's way. Some squads knew their individual tasks really well; others were weaker and required a second attempt on the same lane to receive a "go."

As I was about to leave for Fort Belvoir, there was a training accident. I became swooped up in the mayhem. A trip flare used as an enemy boobytrap had exploded burning a squad member's face, arm, and hand. He required immediate attention. Soldiers are trained to attend to these injuries, and one of his fellow buddies went into overdrive. This man treated the injuries with perfection using what first aid supplies a typical rifle squad has on hand. He personally took charge and instructed the others in how to make a buddy-carry system so that the injured soldier was evacuated to the aid station that had already been alerted to an incoming casualty.

As it turned out, it was the squad leader that carried out the evacuation. I witnessed responsible actions by a trained soldier. It was an extraordinary example of small unit leadership.

I thought about what I had seen on the way to my temporary living quarters, and upon arrival I went straight to my office. I sat down behind the typewriter and wrote a detailed summary (Trip Report) of the day's events including the real life-saving actions of a squad leader. Then, I was compelled to nominate this leader for an award. The next day I turned the trip report and award recommendation over to the chief.

The Army Achievement Award was established to recognize extraordinary service in the line of duty. What I witnessed was nothing short of outstanding. I never heard if the award recommendation submitted for the soldier's quick thinking was approved. I wrote the recommendation for his company commander's consideration as I was not in the squad leader's chain of command.

HACKLES

The day had come and gone. It would be another late evening in my Fort Belvoir, Virginia office. My family was over a hundred miles away in Tidewater, Virginia so it didn't matter if I worked into the night. Besides, it was freezing cold outside. At some point I became so fatigued that I stopped to rest my weary eyes. I pushed away from the desk and stretched out on the office sofa. On the end table sat a small portable radio. I turned it on and tuned in to a local station. A "Coast To Coast" episode was already in progress. Art Bell, the show's host, spoke about Georgetown before a commercial break. Georgetown! Could he be talking about the community in DC just across the Potomac River from Arlington and Falls Church, Virginia?

Following the sponsor's break, Art Bell continued his discussion with a gentleman, his name is lost to memory, on the topic of exorcism, specifically on the Georgetown house seen the movie *The Exorcist*. This really got my attention, and I forgot about my fatigue. I listened in carefully. The movie was filmed inside a New York City home, but the outside work was filmed at Georgetown University. It turns out that the university owned that particular real estate that adjoined the school.

Rolling back a decade or more, I'd seen *The Exorcist*. It was the most haunting and unnerving film I've witnessed. I don't recommend it, and I will not see it again. It's the kind of evil that shakes me to my core. Discarnate entities have always been of interest to me, and I believe in ghosts, but possession of the body is freaky. The movie, though a work of fiction, is based loosely from a documented exorcism that occurred in Saint Louis, Missouri. The home in New York resembled the one in Saint Louis. The movie's director was a Georgetown University graduate, thus the house. It looked shadowy and spooky at night especially dimly lit by street lamps.

The radio program ended leaving me with an eerie feeling about this house in Georgetown. I was a bit unnerved by the memories of the movie. I had explored a little over there and wondered if I had passed by that house. Just then I heard a window slam shut. I thought that I was the only soul in the building. I walked around to see who was present. Finding no one, I felt the hackles begin to raise on my neck and spine. It was time to retire to my temporary quarters located down at the end of the street. I was relieved to find the company of others doing the same thing.

INVITATION

By now I had written a lot of correspondence and reports. The chief must have picked up on my writing style. One Friday afternoon he walked into my transition office and sat down. Then he dropped a bomb on me. "Joe, I want you to prepare a draft invitation to the reactivation ceremony," he announced.

"Great," I said, "who's the addressee?"

"The President," was his reply.

"You want me to write a letter of invitation to President Ronald Reagan?"

"Yes I do, and have the draft on my desk Monday morning," he instructed. With that the chief smiled and wished me a great weekend then he was off.

It was already mid-afternoon, and my wife expected me home in Tidewater Virginia for the weekend. Realizing that my best work was always under pressure, I immediately began working on the draft invitation. I knew the invitation would be transferred in its final format on two-star stationery so I set the margins appropriately on the typewriter. I wrote the initial body of the invitation on my paper tablet. I didn't want it to be in military style so I roughed out a friendly letter into an invitation to fit on general officer stationery. Word selection came easy, the draft was finished in less than an hour. I then ran over to headquarters to procure several sheets of general officer stationery. Returning to the G6, I wrote the draft for the commanding general's signature. It was on the chief's desk as he had instructed before I departed for Tidewater.

The invitation was mailed, but the President didn't attend. The Secretary of the Army, the Honorable John O. March, was his White House representative. A few weeks prior to the activation ceremony, the White House sent a framed letter of appreciation signed of course by Ronald Reagan to the headquarters. Everyone was so proud of his letter acknowledging our efforts to rebuild the famous 29th Infantry Division. His framed reply was placed on the wall in view of everyone entering or leaving the headquarters.

RENOVATION

I had requested military quarters on Fort Belvoir for my family upon arriving for duty in early January 1985. I knew it would take months before suitable quarters for my pay grade (04) would become available. I was really excited when I received a call from Fort Belvoir Post Housing. I went over to discuss details. Post housing anticipated availability as early as May. A five month wait for quarters surprised me. After a telephone conversation with my wife, we put our house on the market. My wife brought two small boys into our union. Both youngsters were in elementary school in Tidewater. When I reported for duty at Fort Belvoir, I instantly became a geographical bachelor. The separation period was difficult on both of us.

We had built a new house in Tidewater. It contained all the modern conveniences of the day. Unfortunately for us, homes like ours were still being constructed in a massive former strawberry patch near the Princess Ann Courthouse. It wasn't a seller's market. After only three years of living in Tidewater, we found a buyer through a realtor. That was the good news. The bad news was that the realtor's commission would rob us of any profit on the sale. We were lucky to recover our original investment. And, we were lucky to be moving onto Fort Belvoir after the children had finished school for the year.

A government contracted moving company packed and loaded our personal property on a moving van trailer. This was a door-to-door move; we were unloaded and unpacked the following day on Fort Belvoir. The quarters provided was on Jadwin Loop. The loop consisted of townhomes — five quarters to a building. Our set of quarters was the fifth unit in the first building. It was within easy walking distance to Williams Hall. That was the good news. The bad news was that these buildings had been constructed in the 1930s. While they were three-bedroom two story homes, none had central air conditioning. We had to purchase window air conditioners to beat the heat and humidity. I overheard one of the movers in passing say to my wife, "Gee lady, you moved from a nice new house into this. Government quarters like this one must be some kind of punishment. What did you do to deserve this?"

In the end she and the boys adjusted to the sound of window air conditioners. Most importantly, we were together under one roof again, and my wife used her talents to make a comfortable home.

Not long afterwards, late in the afternoon, the doorbell rang. A representative from the infamous post engineers was standing by the front door entrance with a clipboard in hand. He asked if he could enter our home. We sat together in our living room to listen to his news. The man told us that all the row houses on Jadwin Loop had been scheduled for renovation. Yes! He went on to say that both upstairs bathrooms would be remodeled, and that the kitchen would be completely remodeled too. It got even better when he told us that a central air conditioning system would be installed. We were hugging each other overjoyed by the news. Then I asked the engineer when do we move into temporary living arrangements?

He looked at us and said, "You're not moving anywhere. We'll install plastic sheets to help cut down on the dust and grime. You'll have to eat out when we gut the kitchen, but that shouldn't take more than a week to ten days. Our men will renovate one bathroom at a time so there'll always be running water available." Our excitement hadn't lasted long. Six weeks later, the job was finished, and we eased back into a more normal lifestyle.

OPPOSITES

A meeting was scheduled in Richmond, Virginia for a discussion on matters of unit reconfiguration and reactivation ceremony details. On a previous visit to Baltimore the division staff had participated in a very successful meeting with the Maryland State Headquarters on the same issues — reconfiguration and reactivation. We met in a conference room at the Virginia State Headquarters. All about the room were huge portraits of Robert E. Lee, Thomas "Stonewall" Jackson and other famous Civil War Virginians, but I thought it was strange that the "father of our country," George Washington, was nowhere to be seen.

Both staffs were seated as opposites at a huge rectangular shaped conference table. Colonel Bilo represented the division staff. In the room there were approximately a dozen senior officers. Colonel Bilo began the meeting and briefed the issues. He would call upon members of his staff for clarification when necessary.

Then the state headquarters staff began to push back on matters without offering solutions adding to the list of unresolved issues. The tension in the room grew making the exchanges awkward and embarrassing. Finally, Colonel Bilo had had enough. He abruptly closed his notebook and shoved it in his satchel. His next action was more astonishing. Standing, he instructed us to pack up saying, "We're done here!" Then we walked out of the room, left state headquarters, got in our staff cars and headed for Fort Belvoir.

I struggled with the differences between attitudes of the two state headquarters. How could one be so amenable and the other so resistant? It left an impression upon me. The 29th Infantry Division (Light) would consist of regiments from Virginia and Maryland just like it had been comprised during WWII. Both states had agreed to the reactivation of the division, and the secretary of the army (SOA), the individual who made the ultimate decision to select a division, was a Virginian and a personal friend of TAG-Virginia. I was confused and disappointed. As a soldier, I tried to put myself in their positions, but could find no answers to their opposite natures. One pushed and the other pulled. One embraced the division concept and the other pushed back on it.

In the end, the generals would work it out, at least they did in Virginia. Maryland was on track. The NGB expected cooperation as did the CSA and the SOA.

CONCEPT

The G6 was responsible for the Division Concept Plan. The new light division was similar in design to the divisions of World War II only fewer in total troop strength. The force's weapon systems since the early forties had vastly improved thus reducing the need for additional troops. The entire 10,600-man light division could in theory be transported by aircraft to just about any location in the world in a matter of hours. My assignment was to compile input from each of our G staff primaries and organize it into a notebook.

It was a monumental undertaking as all the various milestones for conversion to the new light division's table of organization and equipment were scheduled on paper and coordinated through the various commands. Most of these actions were driven by logistics. It required many trips to Baltimore's 5th Regiment Armory (state headquarters) on 29th Division Street and to Richmond's state headquarters building at 501 E. Franklin Street.

The crown jewel for me was the task of writing the executive summary. Though it was written for the chief of staff's signature, I was its author from the transition office. Having read through the documents gathered over the months of planning, and notes taken during meetings at division headquarters, at NGB in the Pentagon and both state headquarters, I was up to speed on the process of converting to the new table of organization and equipment. The 116th Infantry Brigade (Separate) and the 58th Infantry Brigade (Separate) would reorganize into identical light brigades. They would be designated 1st and 3rd Brigades of the division. Division Artillery would mostly be comprised of Virginia units. Division Support Command would essentially be comprised of Maryland units.

The division's Aviation Brigade would come out of both states as well and headquartered in Maryland. Division Troops, i.e., engineers (VA), signal (MD), administration (personnel support - VA), military intelligence and the long-range surveillance (MD), reconnaissance (MD), military police (MD), and the band (VA) would be divided about equally between the states. The monster in the room was the 2nd Brigade.

The brigade headquarters and two battalions would be organized according to psychologist Bruce Tuckman's memorable phrase describing the stages of organizational development, "forming, storming, norming, and performing" just like the division headquarters — from the ground up at Fort Belvoir. One infantry battalion from Maryland, 2nd Battalion 115th Infantry, would round out the 2nd Brigade but would remain under the 3rd Brigade's operational control during peace-time conditions. The executive summary in final format dated May 22, 1985 stretched out to five pages. In my view, the summary was written so practically anyone could understand it. The concept's supporting documentation, i.e., transition and conversion scheduling, manpower, training, readiness and mobilization, logistics, facilities, filled a notebook. The executive summary of the Division Concept Plan is the appendix.

CEREMONY

After nine months of planning and coordination, and the renovation of a new division headquarters at Williams Hall, the day of reactivation had arrived. This was a glorious end for my austere digs since joining the staff on January 4, 1985.

We physically moved into Williams Hall the week of reactivation. The commanding general's second floor office commanded a view of the entire parade field. The morning of Saturday, October 5, 1985 was a typical fall morn. The dew had settled overnight on the static displays making them shimmer as the morning sun rose, burning it away. My duty was the oversight of the reviewing stand. Colonel Bilo, division chief of staff, was the commander of troops.

A six-gun battery of artillery rolled in and positioned themselves in its predesignated spot on Long Parade Ground. Not long after the troop units began arriving in chartered buses along Belvoir Road. Every unit in the light division was represented by a composite group of soldiers. At the far end opposite the firing battery the band began playing John Phillip Souza march music. All the while invited guests, dignitaries, "VIPs" and interested bystanders arrived. The reviewing stand was bustling with activity.

Virginia Governor Charles "Chuck" S. Robb (stepping into the vehicle) and Maryland Governor Harry R. Hughes (looking back) prepare to troop the line. Long Parade Field, Fort Belvoir, Virginia October 5, 1985

No one wanted to sit or stand in the morning sunlight very long. The Honorable John O. Marsh, Secretary of the Army, represented the Department of Defense and was the US President's representative. There were two state governors, loads of "brass" from the Military District of Washington with generals out the wazoo and especially all the troops standing at parade rest on the parade ground in front of the new headquarters. On cue, the commander of troops ordered, "Bring your units to attention!"

A calming emotion washed over me. Nine months of reactivation planning had come down to this very moment. All eyes and ears were fixed upon the array of division units lining the uniformed field. All the regimental colors and battle streamers waved gently in the breeze. Everyone fell silent waiting for the presentation of the division's color-guard. I felt at ease. An inner voice said to me, "This is it, history in the making, enjoy the show."

MONSTER

According to the division's concept plan, headquarters of the 2nd Brigade was located at Fort A. P. Hill, Virginia. It was scheduled to come online with an effective date of July 1, 1986. Colonel John D. Lancaster was selected to command it.

Camp A. P. Hill, Virginia was created during a time of tremendous mobilization and train up for America's entry into World War II. Originally named after Ambrose Powell Hill, a US Military Academy (West Point) graduate, a Virginian, and Confederate Army general, the installation quickly grew to support an influx of soldiers and equipment in preparation for the nation's defense and possible deployment overseas. Individual farms and whole communities were swallowed up through eminent domain to form what today is known as Fort A. P. Hill. This sprawling seventy-seven thousand acre post is located approximately twenty miles south of Fredericksburg, near the town of Bowling Green, and contains adequate terrain to support a variety of training activities.

Reorganizations are common in the military, but they are never really accepted or embraced by those affected. The upheaval is always disconcerting. One day you're an artilleryman and...the next day you're something else, and usually untrained for a new occupation.

For A Battery, 2nd Battalion, 111th Field Artillery, July 1, 1986 began like any other day on Fort A. P. Hill, a hot and humid mid summer's day. By days end, this battery would experience a redesign, complete reorganization and physically move forty miles south of Fort A. P. Hill. Its transformation would increase the unit's strength by thirty-three personnel topping out at a total of one hundred soldiers and replacing its artillery and haul vehicles. In reality, the battery had already moved away and the headquarters of the 2nd Brigade opened for business in the facilities they had occupied. Having been a hot summer's day, the mood had been bitter sweet. It was bitterly hard to move, but sweet because expansion created opportunity.

A Battery's strength was only sixty-seven officers and men. Over the years it had learned to use every available square foot of space in its small compound on Fort A. P. Hill. An infantry brigade headquarters was bigger. A colonel would have to adjust to the battery commander's (captain) old office. Sixteen other officers, one warrant officer, and eighty-two enlisted members would stretch the facilities to more than its maximum capacity. In time, there would literally be standing room only.

Cross-leveling is a military term often used to describe many transitional actions. However, the birth of a brigade and a division headquarters, meant that troop and equipment requisitions were submitted to the next higher headquarters, and regularly monitored. The soldiers of both units were forming a common personality.

Although facilities were grossly inadequate, the lure of new soldiers was irresistible. In a system where there are limits on the number of grades for both officers and enlisted members, the reorganization and expansion of a new headquarters seemly overnight created slots for upward mobility. This made progression possible. Field-grades and senior non-commissioned officers were competing for new slots. It would be a bonanza for individual promotions.

Throughout the remainder of 1986, the brigade headquarters continued to staff up. Its ability to control would take time to reach maturity. It needed individual and collective training plus proper equipping to achieve an acceptable level of readiness, and that required focused attention.

Division training objectives to obtain goals trickled down and out to brigades and separate battalions, however, 2nd Brigade was again unique among brigades. It had grown out of an artillery battery. Most of its personnel traveled south to a new location. Those who remained had to re-qualify in another military occupation. New recruits were quickly sent to basic training and then on to qualify in one of numerous specialties. As for the infantry battalion, it was a tricky situation. Most of the 3rd Battalion, 116th Infantry of the 1st Brigade was reorganized and re-designated as the 1st Battalion, 170th Infantry of the 2nd Brigade. 1st Battalion, 183rd Infantry was more complex. A cavalry troop, a detachment of a support battalion, an infantry combat support company, and a portion of an engineer company were all redesigned to form the companies of the battalion. Personnel shortages were recruited and sent to basic and advanced training. It all took time to convert and jell into an infantry battalion.

Infantrymen of the 2nd Brigade's newly activated infantry battalions continued to train to become individually qualified. Those re-designated companies focused on squad level battle drills. Staffs of each battalion and the brigade headquarters drilled on exercises to sharpen and strengthen their respective command and control measures. 2nd Brigade worked hard at becoming combat ready. 1986 was a year of unprecedented change. It was the monster in the room. In time it would become a highly functional unit.

CHAPTER TWO

G1

The reactivation of the division was a resounding success. Following the pass-in-review ceremony, those attending were invited to an open house... a chance to cross the street and walk through the new division headquarters building, Williams Hall.

Headquarters was crowded with folks for a time, then it began to trail off as they departed.

That evening the command group hosted a formal dining-in at the Fort Belvoir Officers Club. It was a fine affair and a fitting end to a very long day.

FLOODED

Fort Belvoir was no different from most army posts. It had an enormous power plant supplying electricity and hot water heat to all of its facilities. The heat was generated and pumped underground throughout the post. Hot water radiators were turned on at a predesignated date in the fall and then turned off in the spring. Some days it was pretty chilly inside until the sun warmed the outside air.

Such was the situation before the division's reactivation. However, sometime in October, on that predesignated date, the hot water began to flow in the mostly underground pipes. Unknown to those who had recently moved their offices into Williams Hall, the hot water valve in the second-floor ceiling had been left in the open position during the many months of renovation. No one remembered to close the valve when the reconstruction project was completed. Sometime during the night, the hot water began to flow as scheduled. Hot water shot out above the commanding general's office collapsing the ceiling. It then ran out into the hall flooding nearly all the second floor. Water was cascading down the slate treads to the first floor working its way into the lowest point, the basement. The building was soggy the next morning when the first person arrived for work.

The carpets on three floors were mostly ruined as was the ceiling area in the commanding general's office. It took another month to dry out, replace carpet, drywall, and paint before Williams Hall returned back to its original condition.

TRANSFERRED

The ad hoc group known as the G6 continued on beyond the reactivation of the division and well into the next fiscal year. Lieutenant Colonel Zajac remained in position as the G6 and G2. I was transferred.

The Assistant G1 was reassigned thus creating a major's vacancy, landing me in an infantry slot for which I was branch qualified. This was a relief for me, but it wouldn't last long. I was detailed to the task force to implement the light division concept. My assignment was to manage the division's personnel and administration right down to the effective date of July 1, 1986.

I moved lock, stock, and barrel from a very dilapidated WWII building into newly renovated Williams Hall and began to function as the division's Assistant G1 and Deputy G6.

INTENT

As the light division moved into its initial operations and training year, the commanding general (CG) issued his intent. "I want you to focus on what I am convinced must be our number one priority this year — soldier power. Our soldiers are the fighting fiber of this division. Soldier power results from excellence in training, small unit battle drills, marksmanship, mastery of field crafts, physical conditioning, and discipline. Anticipate problems, think ahead, by war gaming and brainstorming with key non-commissioned officers and officers. Use the brief-back technique. I expect every level of command to be briefed by your subordinates on exactly how they plan to execute your training guidance. The bottom line of all this effort is to make certain, do not guess or assume, that every soldier will receive challenging, worthwhile, and productive training and that every light leader is coached, encouraged, and guided by his commander. Develop leadership at all levels, but concentrate on the leadership level that makes direct contact with our soldiers," were the very words used by the CG.(2)

INTEGRITY

In a meeting with the chief of staff, I was briefed that the Secretary of the Army (SOA) requested assistance from the division. I was tasked as the division point of contact.

A request from the SOA to division headquarters may seem a bit unorthodox, but the SOA, a Virginian, was a good friend of TAG-Virginia and he had a vested interest in the chief of staff. It was understandable considering their relationship but not appropriate protocol unless direct coordination had been authorized by the state headquarters. To my knowledge it had not.

The army was in the midst of organizing light infantry divisions and along with it was the restoration of the regimental system. Soon, divisions comprised of brigades were composed of regiments again. As the light divisions were to be "elite," every light leader studied warrior ethos including their regiment's history. The Maryland and Virginia Army National Guards had always maintained unofficial regimental integrity.

Since the history initiative was the SOA's brainchild, he planned to conduct a staff ride through a nearby Civil War battlefield. The use of weather balloons to indicate troop distances and locations better illustrated his teaching points. To demonstrate this idea, he required assistance.

The division artillery had the balloons in its meteorological section. Soon it was on the road traveling to one of Fort Belvoir's training sites. Upon reaching the designated spot, a handful of artillerymen began to employ balloons, a pre-rehearsal prior to the SOA's rehearsal. Immediately problems surfaced. A defective inflating device had to be replaced causing a delay of several days. Then a storm turned the earth into a muddy bog. Foul weather continued ultimately postponing the SOA's rehearsal. By this time rumors were circulating that something unusual was happening at Fort Belvoir.

From the beginning the guidance to me was not to divulge any information on the mission. I received a telephone call from a colonel seeking information on the use of weather balloons. He was a technician for TAG-Virginia.

Specifically, he wanted to know what was going on at division headquarters that involved the use of the division artillery's meteorological section. It was an easy question to answer, yet it placed me in a difficult position. Having been instructed not to discuss this mission, I felt squeezed. If I talked, I violated my instructions and perhaps lose my boss's confidence and loyalty to him. If I didn't talk, repercussions were likely.

I told the colonel that I hoped he understood that I was not at liberty to discuss the mission. This did not bode well, and intimidation ensued. As he pressed, I had no choice but to refer him to my superior. The conversation ended. I remember going on a long run relieving some tension.

Soon after, I was enjoying a peaceful weekend at the state military reservation (Camp Pendleton) on the southern end of Virginia Beach. Around 0600 hours someone knocked at the door. The installation's deputy commander told me to report to the headquarters building immediately, TAG-Virginia, Major General John G. Castles, wanted to speak to me.

Approaching headquarters, I could see the general officer's two-star flag on display. This was no joke! I couldn't imagine what General Castles wanted with me, but I was about to find out.

I saluted him and was given permission to sit. Then a long series of questions ensued. My answers were candid and of course honest. Several hours later I was dismissed. When I returned, I took a long run along the beach... eliminating the tension.

Back at headquarters, the chief asked how my visit with General Castles had gone. Again, I felt the squeeze as I hadn't mentioned our meeting to anyone. "It disrupted my short respite at the beach, chief," was my forthright answer.

The SOA finally conducted his staff ride at Fort Belvoir, and the drama ended as quickly as it had begun. A seemly harmless tasking had tested loyalties, but my integrity had prevailed. Honor carried me through, and I passed the litmus test. Nothing was ever said again about the SOA's staff ride.

BUCK

Division G-staff meetings were common, many lasting well into the late evening hours. One day we conducted a marathon meeting that ended around midnight. Everyone was eager to leave, and I found myself to be the last soul in Williams Hall. After ensuring that the headquarters building was secure, I stepped outside to lock the front entrance. Flanking sconces were on casting bright light on the door. Locking the deadbolt, I turned around only to find a deer (buck) standing at the bottom of the steps. He was a sculpted mature animal with an eight-point rack.

Several of my neighbors were deer hunters and had hunted on post. I remember them talking about a magnificent buck in the woods. It was a common sight to see the doe in yards eating flowers, shrubs and new grass with their young in the late afternoon, but I had never seen a buck.

Stepping down the stairs I figured the buck would bolt. I was wrong. It stood its ground. I backpedaled to the top of the landing and turned to use the parallel set of steps. To my astonishment, the buck pranced over to the steps to meet me again. Something was going on here. I quickly unlocked the front door and stepped inside. I flipped the light switch on and off several times but the buck wasn't deterred.

I left the building through a backdoor entrance then crept around the side of Williams Hall to observe the buck. When I looked, it had vanished into the night as if it had been in my imagination all along. I thought for a moment I had seen a apparition.

Walking to my quarters on Jadwin Loop was a bit creepy. It was a moonless night and made darker by the absence of celestial light. I stopped every now and again looking for the buck. I had the heebie-jeebies that it was right behind me.

SCROUNGER

Upon returning home from France, the division command sergeant major donated a jar of sand to the headquarters. He had been to a Normandy D-Day anniversary as part of a division headquarters' delegation. The sand was imported by him from the "Omaha" Beach sector of Normandy Beach along the French coast.

The chief of staff asked if there was a place of honor in the building for this jar of sacred sand. There were other mementos laying around in various places in Williams Hall. Most of these were donated by surviving D-Day veterans of the division. In those days it was fairly common for a 29'er to suddenly appear at headquarters just to visit and chat about his experiences in WWII. We always took time out of our day to entertain them. The answer was no, there was no designated place for gifts.

Not far from Williams Hall was a post museum. The installation was originally established in 1915 as a rifle range and training camp for troops quartered at Washington Barracks. In 1917 the training area was named Camp Humphreys and later changed in 1935 to Fort Belvoir. *Belvoir* in French translates to "beautiful to see." I visited the museum after making an appointment with its director.

I told the museum director I was on a scavenger hunt. I was searching for a museum quality glass case — one that would be able to support and secure mementos donated by veterans of the division. He didn't have a case by that description to donate, however, he indicated there was some "junk" in the basement I might be interested in. The basement was full of discarded remnants, but I did spot a cabinet that had possibilities. The front glass was missing, presumably broken and removed long ago.

The cabinet stood approximately eight feet in height and was half as wide. It was constructed on a pedestal that had a flat top surface. Four adjustable shelves fit inside the rectangular case. It would fit perfectly in the foyer of Williams Hall I envisioned. All it needed was a plate of glass to fit in the hinged frontside door, and the cabinet would be whole again. I told the director we could use it if I could fix the door.

I briefed the chief on the find. He coordinated the funds needed to purchase glass for the installation of the cabinet's front door. The cabinet was moved to the foyer in Williams Hall and thoroughly cleaned. Within a week or so window specialists measured the aluminum frame door for a plate of glass.

Within a week the cabinet was filled with odds and ends of donated 29th Division soldier personal items and other gifts presented to the division before, during, and after the reactivation ceremony.

BOOKS

A call from the headquarters company conveyed an amazing story. It seems that a tractor trailer loaded with army manuals arrived at Building #2247 — the original division headquarters. I dropped what I was doing and immediately left Williams Hall, driving my car to Building #2247. Sure enough, there sat an 18-wheeler. Its driver explained that his job was to deliver not unload it. He handed me the shipping order. Upon examining the order, I discovered that the previous G1 had ordered all the books, mostly field manuals and regulations, necessary to outfit the division headquarters plus all the subordinate units, i.e. brigades and separate battalions with a complete library of army "how to do anything." After looking at the trailer's load, I determined a squad of men was needed to unload and store the books. But where?

The matter could have been easily managed by listing the subordinate units' identification code on the order. The only code submitted was for division headquarters at Fort Belvoir. It was an oversight, but that didn't matter now as the books were on a trailer, and the driver expected to be released. After some coordination, I convinced the driver to drop his trailer in place for a few days since the load had been assembled in the military district. I needed more time to organize a work detail. Fortunately, he complied.

While searching for a work party, the headquarters company commander suggested that the books could be stored in one of his unused buildings. It was another two-story WWII barracks building near the trailer. That would have to do. By the time the trailer was moved around the corner a work party had arrived. It took all day and part of another to unload. Military books of all types, hard bound and soft, were stored on the first floor of an open bay until it was filled with stacks of paper manuals standing at or above window height. Then, there was no other option but to carry books upstairs to the second floor. It too was nearly filled when the trailer was finally cleared.

With that problem solved, I began to think about how to distribute each unit's library of books. I received another call from the headquarters company. A representative from the post engineer's office had made a routine inspection and discovered the cache of books. He reported that the weight was too great on the building's foundation, especially the second floor's load-bearing capacity. He gave us a couple of weeks to remove the books.

The inspection report was really a blessing in disguise. It elevated the distribution of books to a much higher priority. The G3 was conducting a training meeting at Fort Belvoir the following week. The chief recognized an opportunity to solve several issues. The G3 didn't like the chief's additional guidance for his training meeting.

TRADITION

In the old days, bugle calls were used for a multitude of purposes such as officer's call. In the modern army, radios and the telephone replaced the bugle. The tradition of calls generally fell by the wayside. The commanding general expressed an interest in conducting his first officer's call. The chief of staff tasked me to help think of ways a division headquarters officer's call might set a precedence for a division level tradition. Several of us began to think of ways that might establish a long-standing tradition.

Most of the ideas suggested were patterned after regimental customs and courtesies while others were offered from a regulation that provided policies and procedures for honors, ceremonies, and customs. We came up with a recommendation that each officer of the general staff be issued his or her own drinking cup. It was adopted. With a cup in hand, many types of honors could be rendered at an officer's call.

With the concept approved, the work began to select a cup design. I had been the recipient of a Jefferson Cup long before joining the 29th Infantry Division (Light). I brought it to work one day to support my recommendation for Jefferson Cups.

A Virginian, Thomas Jefferson, commissioned a silversmith to make eight cups from his own design. The sterling silver cups were simple yet versatile. In time, they became very popular. Polished pewter is as beautiful as sterling silver but much less expensive especially when purchased in bulk.

I convinced the group to accept my recommendation for the Jefferson Cup. Each cup was engraved with the division's patch and individually numbered. The numbered cups were registered in a logbook—a cup was assigned to each officer. Upon reassignment, the officer was allowed to take his or her numbered cup as a token of service to the division. A box approximately the size of an army footlocker to store and transport the cups was generously donated.

The headquarters commandant issues Jefferson Cups before the start of an officers' call at the Fort Belvoir Officers Club. Photograph is dated April 5, 1986.

At the first officer's call, individually numbered Jefferson Cups were distributed. The original commanding general was assigned cup number 1. With that, a tradition ensued. My cup number was 11.

BUGLE

Working and residing on an army post is a unique experience. I thoroughly enjoyed the years I lived in government quarters on Fort Belvoir, Virginia. I embraced the culture, customs, and atmosphere that are very different from the civilian world. Living as well as working on a post had advantages and its disadvantages, but in the end my neighbors — the army family — provided me a richer appreciation of military life whether in the training areas or the garrison.

My son, Joel Keith, was born on December 17, 1987. The birth took place in the delivery room at Dewitt Army Community Hospital on Fort Belvoir. (The hospital changed its name in 2007 to Fort Belvoir Community Hospital.) My wife and I had participated in Lamaze classes. I assisted in the birth process. When a nurse brought me the sleeping infant, someone in the pediatric ward had neatly rolled an army wool boot sock into the form of a skullcap and placed it around Joel's head. He had made his first formation in style.

Having been granted the proper security clearance that fall, I was scheduled to depart for the Military Intelligence transition school at Fort Huachuca, Arizona. The school was scheduled to begin in late February. Since we would be separated for approximately four and a half months, my mother-in-law moved into our quarters on Fort Belvoir to assist her daughter in my absence. Fortunately, this arrangement worked out very well, and upon returning home from school, the family began to normalize.

In my absence, my wife and mother-in-law had purchased a baby stroller for Joel. I loved to stroll him in the neighborhood and especially around Long Parade Field where mature hardwood trees provided shade protection from the sun. It was one of my favorite chores on holidays and free time on weekends. I enjoyed talking to him and, well, he babbled back and would often throw the teether on the ground for the pleasure of my retrieval. One day I really got his attention.

It was late in the afternoon and we were strolling around Long Parade Field. At precisely 1700 hours (5 PM) the intercom from post headquarters sounded Retreat — a bugle call to mark the end of the duty day.

Military historians record that when George Washington first assumed command of the Continental Army in 1775, drums and bugles were already a fixture of military life and operations. In a time before radios or texting, there was no other way for a commander to communicate his orders over the din of battle. As the US Army evolved, it standardized the use of bugle calls for a disciplined lifestyle. The calls united soldiers and their families. Bugle calls helped to maintain pride and foster a greater sense of community on installations.

Standing behind the stroller, I knew what was coming. Boom! The cannon blast just about sent Joel into orbit. The intercom began playing *To The Color* — a bugle call representing the national anthem. (As the bugle sounds the notes, the US Flag is lowered to the ground.) All personnel within hearing distance of this cannon shot and bugle call are to show respect for the flag of our country by standing at attention and facing in the direction of the flag. My boy was wailing like a banshee in his stroller over the shock of it all. He certainly had strong lungs as I stood by at attention. After that experience, my situational awareness changed for the better when pushing the baby's stroller around Fort Belvoir.

SGS

As the Assistant G1, administration was almost foreign to me. I could tackle most anything. In the future, I would earn an advanced degree in public administration. However, the sharks were circling, and I felt like I was standing at the end of the plank. One more step and I would have become fish food. Mercifully, the chief moved me into another position within the command group — secretary of the general staff or SGS.

In my opinion, SGS was a much better fit for me. If I couldn't be writing an operation order, plan, annex, appendix, tab or enclosure, writing the minutes for a general staff meeting would have to do until another job became available. As always, I thrust my full attention to detail into every aspect of each meeting. Often the minutes for a single meeting would take more than an entire day to draft. I submitted each draft for the chief's review and approval. I took copious notes and kept my remarks as succinct as possible. The several notebooks of minutes became a compendious resource when revisiting an unresolved issue. When I wasn't writing meeting minutes, I was engaged in actions like matters of protocol for visiting dignitaries and informal or formal investigations.

On one occasion, I was detailed to investigate a missing pistol. The incident was uncovered by a false alarm. A thunderstorm had passed over creating numerous lightning strikes. One hit near the arms room and set off the alarm. The armorer wasn't present when the system was triggered. A standard inventory was conducted before rearming the alarm and securing the arms room again. It was then that the discovery was made. An M1911 was missing from the pistol rack! After interviewing certain individuals, I drew the conclusion that the weapon had been stolen. The theft occurred when the armorer left his weapons racks unsecured. Another soldier detailed to work with the armorer was the main suspect. He, of course, denied any involvement. Unfortunately for the armorer, the daily weapons inventory sheet indicated that every weapon was accounted for. He had purposely falsified the daily inventory in hopes of recovering the pistol from the thief before the truth was discovered. It was a thunderstorm and poor judgment that ultimately cost the armorer his job. As for the missing pistol, the matter was turned over to the Federal Bureau of Investigation.

The SGS is the only position of its kind in an infantry division. As an infantry officer I felt honored to be seated in the presence of the primary G-staff and the general officers at their regular meetings. In many ways I considered myself to be the ultimate insider. As such, I became an advisor on command policy and protocol. I worked very closely with the chief of staff, all the generals' aids, headquarters commandant, and the headquarters company commander.

Though the position of SGS was an insider's dream assignment, it didn't come without its hassles. More than one staff officer attempted to "pinch" me for information.

The job required creativity, flexibility, and loyalty. Effective management and good use of available time were necessary tools to succeed as the SGS.

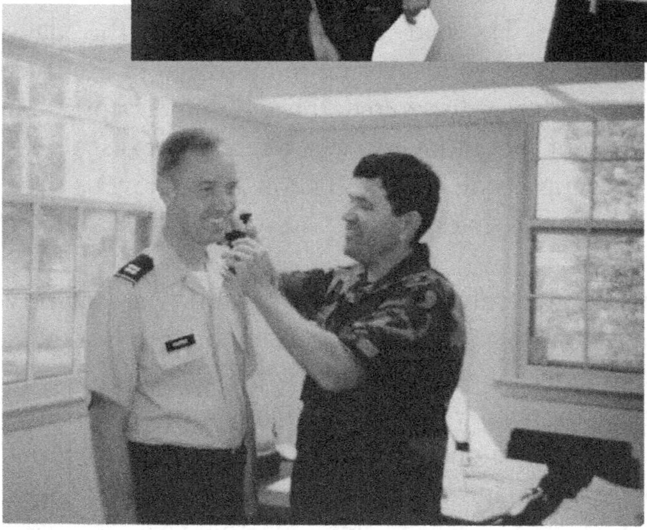

Colonel William C. Bilo removing the author's chevron' s upon promotion to major.
1985, Building #2247, temporary division headquarters at Fort Belvoir, Virginia.

Effective communications, especially staff writing, became a useful vehicle in this assignment. Since time was always a management factor, word selection had to be concise. Information, estimates, and decision papers make or break a staff officer. After attending a class on effective military writing techniques, I've been an advocate of limiting subject matter to one page. "If it couldn't be said in a page, it can't be said," the colonel lamented. This, I adopted as my personal standard — a philosophy I have maintained to the present.

REASSIGNMENT

A new G3 had arrived. The chief reassigned me to be his assistant. I wasn't disappointed by the new assignment. Infantry operations was always in my blood. I was well trained and experienced. It put me in the lineup for promotion. The G3 showed no pleasure at the chief's decision to move me into his domain.

CHAPTER THREE

G3

The light division was in an uncertain position. It had been reactivated, however, to receive its wartime mission, the overall readiness had to rise to meet certain levels. The light infantry training concept had to be implemented quickly. It was. While readiness was in progress, the division created a deception plan.

The division's deception mission was two-fold. First, it had to deny enemy intelligence collectors' accurate information about the actual operational readiness and potential for future operations. Secondly, it had to as quickly as possible to improve unit readiness by individually and collectively training in the use of deception techniques.

The ploy was for all the division units to portray themselves as being less ready than they actually were. Our intelligence collection sources had advised us that many interested folks witnessed the reactivation ceremony then proceeded to waltz through the division headquarters to see what there was to see. This included NATO representatives as well as its adversaries who maintained an embassy presence in Washington, D.C.

To be trained and equipped as light fighters, all infantry squad leaders, platoon leaders, and company commanders, (the key word here is *leader*), would attend a twenty-eight-day Light Leaders Course (LLC) as would artillery and engineer cadre. Once the leaders were qualified, the unit (normally a battalion) would attend a five-day Rites of

Passage (ROP) exercise. The final step in the training concept was to negotiate a twenty-one-day Light Fighters Course (LFC). The rotation of courses continued as long as the light division remained in the force. Attendance at the US Army's Ranger School was encouraged.

AWKWARD

One Monday morning I walked into the headquarters at 0700 as usual. The front door was unlocked, and I could see the chief's car in his parking space. On entering I heard him yell, "Joe, come up here!" The G3 section occupied most of the first floor, so I ran up the steps to the command group. Walking into the chief's office I noticed something on the carpet in front of his desk. "Well, is this some kind of a joke?" He asked.

"What do you mean, sir?" I replied.

"What is that in front of my desk?" He asked.

"Well, sir, it appears to be a pile of crap," was my reply.

"Is it a joke?"

I reached over his desk and took a pencil from his pencil cup; I then pushed the sharpened end into the object. It stuck there just like a miniature flagpole. "It's no joke, chief," I said.

"Who's responsible for this?" He asked.

"It was Mr. Mabe's dog, wasn't it?" He stated.

"Sir, with all respect, Chief (warrant officer) Mabe doesn't own a dog as far as I know," was my reply.

"Well, he works here a lot after hours and on weekends I've noticed. Go find out if it's his dog." Said the chief.

I knew that Mr. Mabe had no dog as I had visited him and his wife, Jan, at their quarters on post. They did have a small boy.

Soon everyone in the headquarters was aware of the massive pile of dog mess in front of the chief's desk, but no one confessed.

Later my telephone rang. It was from the chief. "It's solved, the mystery is solved." He told me the Senior Army Advisor was working in his office for a while and brought his dog with him. Apparently, the Great Dane hadn't finished his business on his way from the colonel's quarters to his office in the Williams Hall. At some point the animal got the urge and slipped out of the colonel's sight and wandered into the chief's office to finish up what had begun an hour or two before entering the building.

The colonel was embarrassed and apologized profusely to the chief. The sad part of this story is that he made his wife clean up the mess. I remember watching from the window as the Army Advisor's wife carried a bucket, a mop, and scrub brushes to the headquarters.

RUN

Jogging was never a fun activity for me ...certainly not a hobby. However, running was part of the army's physical training (PT) test, and all soldiers were expected to meet certain standards based upon gender and age. To ensure those standards were maintained, soldiers were tested every six months. It was called being fit-to-fight.

While stationed in Norfolk, Virginia, I taught courses as part of a program of instruction in the Department of Military Science at Old Dominion University. This was a wonderful environment following the stresses of a command assignment. The cadre had most of the mornings for administrative details, and the instruction generally took place on campus in the afternoons. The assignment was a perfect one to concentrate on your mind and body. One of the captains liked to run. I often joined him in the early morning hours to exercise in preparation for the PT test. This running relationship led to more.

While jogging one day, the captain suggested that we consider training for a marathon. I wasn't sure about his idea and thought it was a bit extreme, but so had been Ranger School. We began meeting daily at the department at 0600 (6 AM) to begin our training. The goal was to run the 1984 Marine Corps Marathon in Washington, DC. The professor of military science at the department gave his blessing so long as we were dressed for duty by 0900 (9 AM) every day.

We began to extend the length of our runs as our endurance developed. It wasn't long before we had built an eight-mile base. On Fridays we pushed further and rested over the weekends. As our goal approached, we ran faster and longer. I began to see changes in my body. I was adapting to the exercise activity. I noticed that my mind would often slip into a neutral zone, and miles would pass before I realized how far I had traveled. My spirit was being fed in a way I had never known.

Running had become a part of my daily routine. I didn't feel complete unless I had worked out. My diet changed too. Carrying a gymnastics bag filled with running gear became normal. I became hooked on running. Running long distances was my physical conditioning and mental acuity when I joined the 29th Infantry Division (Light). Sustainment of that conditioning would become problematic.

The time demands on all the G staff was terrific. We worked long hours by day and often into the night depending on requirements. To help maintain my personal level of fitness, I shifted my running period from early morning to lunchtime with the exception of overnight excursions to Maryland or certain areas in Virginia. I normally worked out first to stretch out muscles and limber joints. Runs varied in length upon terrain familiarity and safety conditions.

I really enjoyed the trail around the nine-hole golf course on Fort Belvoir's south post. (South post is the original side of Fort Belvoir and was bordered by US #1. Today US #1 bisects north and south post.) Constructed around the perimeter of a small golf course, the running and walking track was solid with packed sand and well-engineered bridges that traversed gullies and valleys. In low areas prone to standing water, the engineers built bridging that crossed marshy areas. It was a great trail that meandered through forested hardwoods and rolling hills providing elevation change.

Not far from the historic Mount Vernon, I often wondered when running this track if George Washington was familiar with this beautiful section of ground once owned by Thomas Fairfax. He had to have passed by it on his way to the Pohick Episcopal Church — a church he helped to construct in the first half of the eighteenth century.

SOJOURNER

Staff Sergeant Michael Cooper, administrative assistant for the chief of staff, was waiting. "Sir, twenty dollars for officers and ten dollars for NCOs," he announced. It was time again to sweeten the pot. It was a slush fund for gifts to say our good-byes to folks that had been reassigned. The chief was preparing for another headquarters hail and farewell ceremony.

Searching for a twenty, a slip of paper fell to the floor. My old wallet had accumulated too many reminders of the past. The folded slip of notebook paper showed years of wear, but the handwritten words on it were legible, Brigadier General Archibald Lyon and a telephone number. I had promised the old gentleman that I would contact him when I was settled on Fort Belvoir. Though years had passed I called the general. Here's the story.

His wife, Mary, answered the telephone and summoned Archie. I began with an introduction, but when I mentioned Tony Natalie's name he was all ears. Tony was a US Army retired warrant officer. The two were good friends. I first thought he might have soldiered with Archie, but that wasn't the case. They were brother Masons.

In February of 1982, I had been raised to the sublimed degree of Master Mason. I was fortunate to have completed the degree work in the same lodge where my father belonged. Two months later I reported to Fort Bragg, North Carolina and processed back into the army. I had accepted a job as an assistant professor of military science and was assigned by US Army Cadet Command to Old Dominion University in Norfolk, Virginia. It was at the university's Department of Military Science that I noticed an unusual magazine in the rack. My eye caught the cover. The square and compasses, symbols of Masonry, were embossed on the publication's cover.

I knew from my Masonic work that I had been brought into a deeply rich fraternity... a culture to make good men better. Masonry involvement would have to wait until I finished my army journey, I thought.

Since its humble beginning, the purpose of the National Sojourners has been to affirm the responsibilities military officers have to their brothers-in-arms, the Masonic Fraternity and to our nation. The National Sojourners to this day organize commissioned officers, warrant officers, and Senior Non-Commissioned Officers of the uniformed services of the US (past and present), who are recognized Masons, into chapters to actively provide good fellowship among its members, to assist those who may be overtaken by adversity or affliction, to cultivate Masonic ideals, to support patriotic activities in Masonry and to develop true patriotism and Americanism throughout the nation. National Sojourners, Inc., formally brings together Masonic leaders of the uniformed services of the US into a united effort to further the national defense, and work to oppose any influence calculated to weaken our national security.

The local Sojourner chapter (4) presented an annual award to deserving ROTC cadets. The point of contact in Tidewater, Virginia was none other than Tony Natalie. Contacting him first by telephone, the conversation was positive. I'm sure he was overjoyed to be talking to a recently made Mason from North Carolina. He invited me to visit him at his home in Virginia Beach.

Tony was very enthusiastic about our meeting and shared much of his vast Masonic knowledge with me. Through Tony I learned the history of the National Sojourners—group of military Masons that are dedicated to patriotism and Americanism. I revealed to him that soon I would be redirected to Fort Belvoir, Virginia. It was then he wrote down on a piece of notebook paper the name and telephone number of General Archibald Lyon. Tony encouraged me to contact the general once I was settled in Northern Virginia. I placed the slip of paper in my wallet.

The general invited my wife and me to a chapter meeting in Alexandria, Virginia. He and his wife, Mary, would be our hosts. We met the Lyons at the historic Gadsby's Tavern for an evening dinner and chapter meeting.

The Gadsby's Tavern had been on our list of early American restaurants to visit in Alexandria. We were excited to be combining our first dinner visit with a military Masonic organization. An Englishman, John Gadsby, opened his tavern for business in Alexandria around 1785. The site was the center of Alexandria's economic, political and social life. Notables that had visited Gadsby's Tavern included George Washington, Thomas Jefferson, and John Adams.

We ate a wonderful dinner in an historic colonial setting followed by a meeting of the Sojourners. We were recognized as Mary and Archie's guests. At the conclusion of the meeting, I was given an application for membership.

I eventually applied for membership in the Sojourners and was admitted into Old Dominion #364 in Alexandria. My Sojourner membership in perpetuity (life) card was signed by Colonel Carl L. Sitter, USMC (Ret), a recipient of the Medal of Honor for his leadership in the Korean War (the Frozen Chosin). Old Dominion would be the first of many Sojourner chapters I would join as I traveled the country, and it all started with a point of contact given to me on a slip of paper from a Mason in Tidewater, Virginia.

FAUX PAS

In order to implement the light division training concept, the 29th Infantry Division (Light) was authorized to recruit a fifty-five-man training detachment. It was named appropriately the Infantry Training Detachment (ITD). All the non-commissioned officers were qualified through Fort Benning's LLC. Several were already Ranger qualified. The detachment was commanded by a lieutenant colonel, and the detachment was stationed at Fort A. P. Hill — a training post located in central Virginia.

These men were highly skilled and experienced infantrymen for the most part. Several of them had engineer backgrounds. The ITD had its own staff and compound. All together the detachment operated the LLC, ROP, LFC and several smaller schools to sustain the training of a light infantry division. Notably, it operated its own sniper school.

The chief received a call from TAG-Virginia stating that the American Society for the Prevention of Cruelty to Animals was complaining about horrifying demonstrations going on at Fort A. P. Hill by members the ITD. I was dispatched from headquarters to investigate the claim.

A support battalion (non-infantry) had been engaged in the five-day ROP course. Six hours of the program of instruction were related to field craft techniques for non-infantry units (one hour for infantry units).

Within this portion of the course, squads were required to kill a chicken and prepare it for human consumption. The cadre from the ITD demonstrated how to do this before giving a live chicken to each squad. It seems that Sergeant First Class Nelson of the ITD became overzealous on the platform before his class. He bit the neck of the chicken until he broke it, then proceeded to sling the bird around by its head severing it from the body. The headless chicken then ran around the group spurting blood until it dropped over dead.

Someone was so mortified by the demonstration technique that they felt compelled to report what was observed to the ASPCA. The ASPCA complained to the newspapers. Local news reporters and the ASPCA began bombarding the state headquarters for answers about animal cruelty claims.

In the end, the practice of killing a chicken was modified. An axe was substituted for teeth, and the rites of passage into the light infantry mindset hardly skipped a beat.

A 1988 group photograph of the G2 Section following the completion of its Rite of Passage at Fort A. P. Hill, Virginia

ARTWORK

Living in government quarters is tricky. You never knew what you're going to find available or when. We lived on Jadwin Loop in a five-unit row building. Each unit was a townhouse design. It included a finished basement. The two end units were the most desirable. We became acquainted with a number of families in our building during the five years we resided on Fort Belvoir. It really worked well for me to be within a short walk to Williams Hall.

The Loveladys lived in our building. Rex was assigned to the Engineer school as an infantry advisor. They were fine Texas folks, and we socialized with them often. Rex had a hobby. He enjoyed working with stained glass. Their quarters were filled with attractive free-hanging stained glass windows. He was creative and skilled.

One day I asked him to consider making something for me. He agreed, but said he wanted to show me how to cut glass and solder the pieces together. When I told him what I wanted, he grinned. Only an infantryman would understand.

Williams Hall was a huge step up when the Division Council selected it for the 29th Infantry Division's headquarters. The headquarters facility had improved from its meager beginnings in a WWII barracks (bldg. #2247) to a handsome all brick building right next to post headquarters. A long sidewalk stretched from the street to the dual staircases with a landing at the main entrance. A single plain glass door led into a foyer on the main floor. Above the door was another plain glass transom. My idea was to embellish the glass transom with a work of military art done in stained glass. Rex said he could build almost anything. I took him at his word.

We walked over to Williams Hall to take measurements. He said, "It was very doable, now let's discuss design." We both agreed that infantry crossed rifles would fit nicely in the transom. We also decided to place the "blue and gray" division patch in the center about where the rifles crossed. We needed stained glass.

We wound up purchasing what we needed from a store in Silver Spring, Maryland. Since we wouldn't find a gold-colored glass we settled on a yellow tone glass to represent brass. From this sheet I would cut out the pieces to build the crossed rifles. Gray and blue sheets were purchased as was a sheet of clear glass for background support. The glass was translucent. The hardest part of this process was carrying the sheets of glass back over the river to Fort Belvoir without breakage.

We went to work almost immediately on the transom project. Rex showed me etching techniques and how to break glass into pieces. He set me loose learning technique on scrap pieces of glass. Using our measurements, we drew out a pattern on poster board. Meeting after dinner in his basement, it was at times very tedious work. However, we eventually cut the pieces to fit the pattern. The design was taking shape!

With all the stained-glass pieces cut, it was time to begin soldering them together. This was a slower process than I had imagined. Melting lead and tin into spaces between cut glass was slow going. A metal frame was constructed to hold the artwork. The outer edges were slowly soldered to the frame. When it was finished, I was surprised by its weight. Rex was worried about it collapsing so we reinforced all the pieces with additional solder and constructed a wooden frame to encase the metal frame. We attached brackets to the outer frame to secure the stained-glass art into the transom. It worked as planned.

We decided it would be best to set the art in the transom while the building was unoccupied. Installment had its own challenges as it required working on a ladder. We both needed to use ladders to lift the stained-glass frame into the transom. It took us well over an hour to set the artwork into place. We then secured and cleaned it.

It was a piece of military art and a gift to the division headquarters. I don't recall asking for permission.

One day a G4 representative asked about the stained-glass artwork. I said it looks great and enhances the main entrance. It should be put on the headquarters' property book. That's what the G4 representative did.

Rex liked Sam Adams Boston Lager. I purchased a case and gave it to him as my appreciation for showing me how to work with stained glass. Now, when I open a bottle of Sam Adams, I toast Rex. Follow Me!

DETERMINATION

The G3 was given the widest span of control of all the staff sections. Along with control came tremendous responsibility. As the Assistant G3 (training), I managed a myriad of tasks. One of those tasks was managing the positions for all service schools. Schools included learning a wide array of unique skills and qualifying courses from everything mechanical to clerical skill sets. It was a job requiring great management and attention to detail. I endlessly stayed on the telephone regarding the allotment of slots for schooling. The G3 also maintained three-inch binders for every service school. An entire wall was taken by shelves upon shelves of notebooks filled with details of every assigned seat to a brigade or separate battalion. In those days, it was the only way to recall who had been assigned a position.

The light division depended on helicopters for mobility, therefore, the air assault course was desired by infantry leaders. Rigging loads for helicopter sling lifts was a required skill for the movement of essential equipment, e.g. artillery pieces, large and heavy quantities of ammunition and other logistical supplies. All of the school seats were issued from the G3. Often the demand for this school was greater than the supply of allocated seats.

Fortunately for the division, the Military District of Washington (MDW), DC operated a regional Air Assault School at the Davidson Army Airfield, several miles from headquarters. A standard two-week program of instruction included rappelling from a hovering helicopter, rigging wheeled vehicles, artillery and pallets for sling carry and an abundance of physical training. The school primarily supported the 3rd Infantry (Old Guard) Regiment and soldiers from the elite 29th Infantry Division (Light).

As the school's assignment officer for the G3, I interfaced closely with MDW Air Assault School leadership. The Infantry Training Detachment (ITD)—trainers of the light division—operated a Repel Master Course at Fort A. P. Hill, Virginia. All the repel master instructors at the ITD were air assault qualified. Therefore, when the MDW school needed instructor assistance, the ITD often supported it by detailing an instructor or two to help out when the course was in progress. It was a healthy working relationship.

Each two-week course began early in the morning with a qualifying physical fitness test. This test was standard for all soldiers and was not very challenging. However, since repelling from a hovering platform required a level of self-confidence and personal courage, fitness tests required each student to successfully negotiate a confidence course. Climbing up ropes and high inverted ladders far above the ground was challenging for the uninitiated. Many students washed out without ever beginning the two-week air assault program of instruction.

My hope was to eventually be rewarded for my work as a G-staff officer. To command a battalion would be a justified assignment. In my mind, I needed to be as qualified for command as I could be.

Recalling my early years as an Airborne Ranger small unit leader, skill sets were necessary to be a stronger officer. Attending the air assault school would enhance my understanding of light infantry air mobility operations, but the G3 said he couldn't afford to let me go to school. I was disappointed by his decision.

I knew from my visits with the nearby air assault school that there was a good chance several seats would be available due to failures on the physical fitness tests. I had seen this occur so many times that I generally filled them with standby substitutes from the light division. Our standby students ensured that each course was filled to capacity. I requested two weeks of ordinary leave, and the G3 approved my short vacation.

Saturday was the course's qualifying morning. The students reported at 0500. I was there to qualify. It was dark, and the only lights were the high beams from several wheeled vehicles aimed at the obstacles. A light frost was visible. I was concerned about the slick conditions. High beams of light and a light coating of ice provided a fear factor for many, me included. Nevertheless, these were the added risks when we began the confidence course. Several prospective students failed to climb a rope, and others wavered on the inverted ladder. Fortunately, I breezed through the obstacles without incident and secured a school seat in the class. The first day of training was scheduled to begin on the next Monday morning. After qualifying for air assault school, I returned to my Fort Belvoir, Virginia quarters.

During my wait for the school to begin, my stepson asked for a bow and arrow set. Rather than purchasing a set at the post exchange, my wife and I thought it would be better to make a bow and arrows from wood saplings. On Sunday we walked into the woods behind our home to cut a sapling suitable to make a bow. A light snow had fallen the previous night. I grabbed a short handle axe from the garage on the way. An inch of fresh snow blanketed the ground. Soon I was chopping away. We could trim the bark away and shape the wood into a bow in the garage. On the return trip, I stepped on a plywood board that was hidden beneath the snow. My foot slipped on it, and down I went. I fell so fast that I didn't feel my hand being sliced open by the heel of the axe's sharp bit. The flesh behind my right thumb was sliced open like a hotdog bun. At first, I didn't feel any pain. The blood was profusely flowing down and dripping off my fingertips. I pressed the cut together with my left hand leaving a blood trail in the snow all the way to the house. It would require a doctor's attention.

In the emergency room (ER) a physician assistant disinfected and sewed the wound closed. It took six sutures. He cautioned me not to use it to allow the tissue to knit together… if it became infected to return to the ER. I didn't mention air assault school.

Assessing the situation, I had a decision to make. Could I continue with the course that started in the morning while I was on leave from work? I decided to try.

My hand was bandaged to prevent infection. It would be obvious to the air assault instructors that I was injured. I was determined to try. Would the stitching hold without tearing? I had no way of knowing.

The course began in the darkness. No one noticed the bandaged hand at first. We performed the physical exercises and ran a few miles. Later that morning I was asked about the hand. I said it was my choice to continue. The answer was, "We'll soon see if it prevents your ability to work with ropes."

In those days, the fast rope method was used by special operators for insertions. The air assault course would use the standard method that required a rope passing through a D-ring or carabiner that was part of a Swiss seat system. To brake, one had to squeeze the rope behind the waist at the small of the back. My brake hand was my injured right, impossible with six sutures. They would be immediately ripped out causing a big problem. To manage the situation, I would have to adjust to using the left hand as the brake. That's what I did.

The course went according to schedule. I made the adjustment necessary though very awkward at first. Using the left hand to brake became the new normal. The only issue for me was pushups. We were constantly being harassed. The punishment was usually ten, sometimes twenty pushups. I had to use the outer edge of the right hand to keep from stressing the stitches behind the thumb. It worked out.

I kept the wound clean and covered for the duration of the course. At the end on graduation day, I was free of the bandage. My salutes and handshakes were nearly perfect.

I returned to duty on schedule having skinned the cat.

CHARTER

It was not unusual for veterans of the division to visit their headquarters on Fort Belvoir, Virginia. I had the pleasure of meeting many a D-Day veteran during my years there. It was by an encounter with Sergeant Bob Slaughter that I learned about the 29th Division Association. I had been a member of another division association, an active division — the 25th Division Association. I was curious about the operations of a National Guard division association. I learned that the famous 29th Division had a very well-organized association. Its membership was mostly WWII veterans. Its posts were located in Virginia and Maryland. There were posts in other states as well. On one occasion I attended a national convention.

At that national convention I was encouraged by many members of the association to petition to charter a chapter at the new division headquarters. I did. It was named Post 84 to denote the year that the provisional division headquarters began to form at Fort Belvoir.

The division chaplain, The Reverend Doctor (Colonel) Richard L. Stinson, was then the rector at the St. James' Episcopal Church at Mount Vernon, Virginia. Colonel Stinson graciously invited the petitioning members to use the parish's friendship hall for a covered dish charter ceremony. I invited Major General (Ret) Bruce Jacobs to be our guest speaker.(3) He accepted and made a terrific impact on the ceremony.

DISDAIN

Over the course of my military career, I was very blessed to work for and with some very talented officers. Many were brilliant, and those less capable compensated by working harder and longer. Most were dedicated and loyal workers. I endured the best of times and the worst of times with bosses.

I can think of two officers that I carried ill will for over my years of service. One man I worked for probably suffered from undiagnosed Post Traumatic Stress Disorder or more commonly known as PTSD. This individual was a battlefield commissioned officer. He was one of a handful of survivors from a North Vietnamese Army attack. As the highest-ranking non-commissioned officer left alive, the gold bar of a second lieutenant was pinned on his fatigue jacket on the very hill where the action occurred. He was bright, but he was not well. He suffered bouts of depression and often would isolate in his office. This officer needed psychological help. I survived his persecution, however, he rewarded me with strong efficiency reports. I didn't admire him and struggled to show the respect due to a superior officer. Looking back, I believe he may have prepared me for a future situation.

The other officer I had tremendous disdain for was also bright and possessed great potential, but his character flaw ended his career. The G3 was upset by the chief's decision to reassign me into his section. Our relationship never evolved into one of mutual respect. After his extramarital affair with an enlisted member of the section was discovered, I was embarrassed by his poor judgment and weak self-control. An investigation pursued. The young married female soldier was transferred. The G3 was given a reprimand and was allowed to remain at the headquarters with a second chance. What little respect I had for the man faded never to return. The morale in the section, I thought, was low. This individual's proclivity for domination never stopped with the warning. In time he was discovered in another web of infidelity. His tenure at the headquarters was terminated. I never saw him again.

L5

Christmas of 1987 was unusual. The home decorations were in cozy little places. A beautiful fir Christmas tree was placed in the living area of our quarters. At 451-E Jadwin Loop running cedar wound around the front entrance banister with a large wreath on the door. Our newborn was exactly a week old.

The staff at division headquarters was enjoying a half-day schedule as Christmas Eve approached. On my way home I stopped by Williams Hall to drop off a document. The car was parked when it happened. Opening the door and with one foot on the pavement, I must have twisted my spine causing a very sharp pain in my lower back. It was one of those instances when you know something isn't right. I got out of my vehicle and walked down the sidewalk up the staircase to the main floor entrance then up another flight of steps to the command group. Once I finished my drop-off, the afternoon was planned for last minute shopping for Christmas gifts. My mother-in-law was a house guest.

I said my good byes and departed. I could tell immediately that stepping down the staircase was more challenging. I held on to the banister all the way to the bottom landing. By then it was all I could do to step unassisted to the front door. I made it. The outside staircase was about the same. However, this time I took it one step at a time leading with the same foot. I'm not sure if I could have made it without the railing. Ahead of me lay the daunting sidewalk to the vehicle. There was no railing on which I had quickly become dependent. I limped to the vehicle; the pain was excruciating.

This was just the beginning of my ordeal. I had to bend and turn my waist to be seated behind the steering wheel. By now I was groaning and moaning. Every movement sent shooting pain across my body, especially in my legs. I was not sure how I would make it into our quarters. At this point I wanted to take the weight off my spine ... anything to make it stop hurting.

Our government quarters on Jadwin Loop were old style units. Five townhouses created a row. The designation 451 was the first row. We were given the fifth (end) unit, 451-E. The units were constructed over a basement resulting in steps to access the front or rear entrances. Stepping up would require a herculean effort.

Exiting the vehicle meant a repeat of the movements that created the injury in the first place. I gritted my teeth and grunted. Once inside I fell back upon the couch. Climbing to the bedroom was out of the question. This couch was my resting place for the next three days!

I stayed on that couch until the day after Christmas. It wasn't the merriest of seasons for me, but as I would learn, the swelling had to ease before standing upright. When I did attempt to stand, it took all my arm strength pushing on my thighs to lift my body over the hips. I somehow managed to clean up. My wife drove me over to the clinic. The doctor I saw said, "Take a couple of aspirins."

His suggestion worked. I guess I was expecting more sympathy. The pain was diminishing, but I was worried about permanent damage. Not moving for three days was disconcerting. Now I was slowly healing. Soon I would be normal again.

Fast forwarding into the '90s, I had a similar experience. This time the doctor ordered a CT scan. The X-rays revealed bulging disc material in my lower back, L5. The specialist gave me my options. One of those options was physical therapy. It was suggested that I tighten my abdominal muscles. (Later the army determined that too many sit-ups was actually bad for your lower spine.) The strategy was that strength would prevent a bulging disc from herniating thus preventing back surgery.

As a side note, the back injury was a culmination of many factors, mostly service connected. Once the disc is compacted, it'll continue to deteriorate and may result in surgery. The documentation was enough to award a certain level of disability from the Veterans Affairs.

CHAPTER FOUR

G2

After a year and a half of security clearance processing, I was finally granted a Top Secret (TS) that was required for formal military intelligence training. The US Army Intelligence Center at Fort Huachuca is in a remote area of Arizona near the International Border with Mexico.

The Assistant G2 was the position I had originally been hired to fill in late 1984. After all the assignments at the division headquarters including the G6/G2, G1 and G3, I wanted to become a military intelligence (MI) branch qualified officer. I received my reward in 1988.

The instruction was the equivalent to the MI Advance Course. It was a four- month residency or temporary duty assignment. The training was much more difficult than I had imagined and further compounded by the burden of the additional duty as class leader.

The group was comprised of young reserve active-duty captains, mostly males from the combat arms, that had been forced into the branch. The majority of them were not happy about the forced move. For them, it was branch transfer or leave the army. Their morale was affected by the army's needs to enter the Military Intelligence Corps. I wanted to be here, and they didn't.

I struggled with their collective attitudes, but all in all they slowly adjusted to a change in career paths. The group members contributed in class as expected and partied hard after duty hours. In fact, their shenanigans at the officers' club finally caught the attention of the school's commandant, and I was ordered to visit the post commander.

"Major Harris," he said, "We've got a problem — I think. Are you aware of the parties in the officers' club each night?"

"Well, sir, yes, I am. Isn't that what the club's all about?"

The general and school commandant both agreed. Then the general began to drill down on why I was there. "The club manager filed a complaint about the behavior of your class," he told me. He went on to describe the attention being created by the class.

"Are you aware of the quarter plop dance," he asked me.

"Yes sir! It's fun and very entertaining considering that we're out here in the desert environment with not much to do after duty hours."

The general asked me to show him the quarter plop. The commandant handed me a quarter. I asked for an empty cup which was provided to me. Before I demonstrate, I said, "This works better with loud music and a few beers under the belt."

Both officers looked at one another. I placed the cup on the floor and walked a few paces away. Carefully, I placed the coin between my buttocks and ...

"We've seen enough Major Harris," he laughed.

"The ruckus in the club has to change somehow," he spouted.

"The females wouldn't appreciate this sort of behavior," he said.

"Well, sir, they, as a group, have better aim than the men," was my reply.

"What?" The general shouted.

"From my observations, the females are better quarter ploppers than the males, sir," I said.

Beyond graduation from the course, I kept in touch with many of these aspiring younger men and women. Many were destined to become intelligence officers of combat arms units. Many of them were deployed with their battalions to Saudi Arabia for Operation Desert Shield — the code name for the US buildup of forces in 1990. Most were engaged with their units during the first Gulf War (1991) in response to Iraq's invasion and annexation of Kuwait. As a class, they were hard working students that liked to have fun when off duty. The quarter plop was their creation, not mine.

BUSTED

The handwritten message, "Call the chief immediately" caused adrenalin to shoot through my body. I waited for the chief to answer his phone.

"Joe, I've got good news for you."

... another opportunity to excel, I thought.

"You're next on my list to be an investigating officer. Come over to the office and let's discuss the details," was his instruction.

Handing me the written appointment, the chief verbally described the situation. A company first sergeant had been accused of tampering with his own military personnel file. The unit clerk stated in his affidavit to the inspector general that the first sergeant told him to add certain claims to his military history. He also claimed that this individual was wearing the Ranger tab among other skill badges that were not documented. The chief handed the first sergeant's personnel file to me.

The accused, Sergeant First Class Gung-ho (pseudonym), was a medical company's senior non-commissioned officer and acting company first sergeant. The company supported an infantry brigade. I studied his personnel file in great detail and immediately noticed an entry that stood out.

A meeting was conducted with the acting first sergeant. I deemed it unnecessary to meet the accuser after a review of the acting first sergeant's file. I had read his affidavit that initiated the investigation in the first place. We met at a neutral site. I explained my purpose and responsibility in this investigation. Gung-ho was visibly shaken. I told him to speak the truth. He verified his initial enlistment date and assignment at the Tripler Army Hospital in Honolulu, Hawaii. I asked him about his involvement in Operation Babylift.(5) Gung-ho stated that he did not participate, but that he was aware of the crash of an Air Force cargo plane. When I revealed his personnel file — indicating he had been detailed to support the operation — he quickly admitted that he told the clerk to include Operation Babylift on his record while assigned to the Pacific army hospital in 1975. I then revealed to him that I was posted at Schofield Barracks on the date of the crash, and that I could prove he was never detailed to support that operation and thanked him for his candor.

Next, I inquired about his skill identifiers. I asked the acting first sergeant to name his Ranger School class designation. He again quickly admitted that he had never volunteered for ranger training. When asked why would he pretend, Gung-ho just shrugged his shoulders and stated, "It was my pride that got the better of me."

Following the interview, I summarized my findings in writing and recommended that the acting first sergeant be removed from a leadership position. I presented the report to the chief.

Later I learned that the commanding general decided to reduce SFC Gung-ho to staff sergeant and removed him from the division.

The Army is a small world. Six months later I was in a urologist's office for a cystoscope. Laying on the cool examination table and feeling very apprehensive, Staff Sergeant Gung-ho walked in to prepare me for the procedure! Woe was my fate.

BOSTON

The 29th Infantry Division (Light) was blessed by its rich military history from participation in two world wars and everything in between them. The division enjoyed an active 29th Division Association made up primarily of 29ers. Now the division had been reorganized and was in the process of maturing into the light division. The two-state division, being separated by the District of Columbia, provided for another unique opportunity. All Army National Guard divisions were augmented by the US Army Reserve (USAR) except for the 29th ID (L).

Augmentation came in the form of military intelligence. The Washington Capitol Area was home for most of the nation's modern intelligence collecting. Many of the civilians working at these agencies acquired their skills in the military. They also brought their work week skills to the National Guard and USAR on weekends. For this reason, the National Guard Bureau authorized the Maryland Army National Guard to organize a military intelligence battalion. The 629th Military Intelligence Battalion (Combat Electronic Warfare Intelligence or CEWI) was organic to the 29th ID (L). This arrangement allowed for excellent training between both states. It also provided outstanding support to the G2 and its ability to provide information to the commanding general. We were very fortunate to have our own military intelligence battalion.

Good fortune again fell my way as the G2 and I were invited to a reserve component military intelligence conference. Here we met with our peers from other national guard divisions. The event was hosted by the 26th Infantry Division in Boston, Massachusetts. The highlight of the conference for me was our bus ride from the conference hotel to Faneuil Hall!

An evening banquet was conducted at the Ancient and Honorable Artillery Company of Massachusetts. This unit was chartered in 1638! Today the company serves as Honor Guard to the Governor of Massachusetts. However, over a hundred years before the American Revolution it was established to train officers from volunteer militia companies across the colony of Massachusetts. Since 1746, the headquarters has been located in Faneuil Hall.

We stepped on a bus from the hotel that would carry us across Boston to historic Faneuil Hall. This was to be no ordinary bus ride. As soon as the vehicle pulled away from the curb it was surrounded by a police motorcycle escort. The bus driver followed directly behind the lead two motorcycles. With lights flashing and sirens blasting the bus driver maneuvered to the middle of the street. The cops had stopped all traffic at the intersections allowing our speed to significantly increase. I remember two motorcycles on each side of the bus. I had never experienced anything like this before. Five minutes later we stopped in front of Faneuil Hall. Stepping off the bus, the first object I saw was a bronze statue of Samuel Adams. Behind the statue was the entrance to Faneuil Hall. The banquet was on the fourth floor.

The visit to the Ancient and Honorable Artillery Company was unforgettable. You could feel the early American history inside this hand-built structure. The company consisted of an armory, library, offices, quartermaster (supply), commissary (food) and of course a quality museum.

PLANS

The headquarters of the 29th Infantry Division (Light) was blessed to be located at Fort Belvoir, Virginia. All of Fort

Belvoir is contained inside of Fairfax County; the county is part of the Washington, DC metropolitan area. The "Metro" is home to many intelligence collection agencies. More importantly it is home to two independent intelligence agencies — the Office of the Director of National Intelligence and the Central Intelligence Agency. These agencies employed thousands of federal employees. Many of them learned their specialty from the military. All of the employees held high level clearances. For these reasons, it was decided to let the state recruit its own organic military intelligence battalion that was stationed just across the Potomac River in Maryland.

The G2 section, stationed in Virginia at Fort Belvoir was the largest in terms of personnel of the G-staff. It was comprised of an all-source production section, a collection management and dissemination section, a counter-intelligence analysis section, and an electronic warfare section. A field artillery intelligence officer was included for liaison between corps artillery and G2 identified high-value targets. When combined with an attached Air Force Weather Team and Engineer Terrain Detachment, the staff swelled to twice its normal size. In essence it worked like a gigantic clock with many turning gears synchronized to turn credible information into tactical intelligence.

My experience with intelligence matters prior to being assigned to the division headquarters had been limited. As a rifle platoon leader, the only intelligence I received was from the company commander, written or oral. As a rifle company commander, I received briefings from a battalion's intelligence officer. Weather and terrain was their message. Little was known about the enemy's whereabouts or its capability. Stepping into the world of tactical intelligence at the division level was an eye-opener.

With assets available like the Long-Range Surveillance Detachment, eyes could be placed upon a threat deep behind enemy lines. Information flowing in from a reconnaissance squadron and especially lateral cooperation from a supporting military intelligence battalion quadrupled the size of the intelligence pouring down the funnel to the G2. I was simply amazed at the amount of energy put into an order of battle analysis and the information disseminated in an intelligence estimate.

Trained, and in most cases experienced, men and women performed the analysis. The Washington Capitol area was a rich environment in the intelligence community. It was my great fortune to soldier with these folks. Their overall intellect instilled an atmosphere of cooperation and teamwork. I never witnessed nor did I perceive an attitude of competition among the members of the section. They were happy campers that took pride in their roles to collectively paint a picture of the enemy's locations and intentions. Our brigade and separate battalion commanders were given credible intelligence. My concern was the trickle-down effect. Would the brigade and battalion intelligence sections dilute the information they had been provided?

As an assistant G2, I concentrated mainly on intelligence preparation of the battlefield (IPB). Sometimes I helped out in operations, but working the plans section for the next operation was my functional responsibility. In doing so I recommended that the G2 consider deploying the section to Fort Huachuca, Arizona for a training exercise. He approved, and the 1989 summer deployment to Fort Huachuca was not only productive, it was also timely.

G2 Section, 1989 at Fort Huachuca, Arizona bound for Andrews Air Force Base, Maryland

CUFFLINKS

While an undergraduate and an Army ROTC cadet, I was fortunate to be inducted into the National Society of Scabbard and Blade—a military academic honors society. I was proud of the recognition, and after years of volunteer service, I was elected the society's national president. In 1989 the staff chose to present its distinguished service award to General Colin L. Powell (6), Chairman of the Joint Chiefs of Staff (CJCS). As the national president, it was my duty to present the society's highest award.

I was familiar with the Pentagon. I had visited the iconic facility on numerous occasions, the first being for the reactivation of the 29th Infantry Division (Light). This visit was different. I was headed to the office of all military offices—the office of the nation's highest-ranking officer, the CJCS. An appointment had been arranged months before my visit. I had been vetted and researched for the general's information. The walk from the parking lot and the various halls of the facility had been a good workout.

Somehow, I wasn't intimidated by the entrance to the chairman's office. I was in awe of the military art immediately outside his door. As I knocked on the door, I remembered that the door handle was as impressive as the wooden paneled doors themselves. The entrance was designed to impress. Then it dawned on me. General Powell's office wouldn't be accessible from the hallway. I grabbed the massive brass handle and pressing with my thumb, the latch sprung open.

A full colonel stood in my view. He waved me in. The colonel was the chairman's secretary. "You must be a representative of the Scabbard and Blade," he said. Since there were several others expected to attend the presentation, I introduced myself. "Good," he said, "maybe you can tell me a little about its origins."

Looking around the room I saw what I thought was the entrance to the chairman's actual office. As I began my spiel that Scabbard and Blade is a college military honor society founded at the University of Wisconsin in 1904, the door opened, and a general officer and several civilians exited the office and passed by me on their way out. As they departed the other two members of my party stepped aside to allow them to exit.

Once my fellow society members were inside, the colonel quickly escorted us into General Powell's office. He was sitting in his padded chair and stood to greet us. We saluted the highest-ranking officer in the entire US military establishment — General Colin L. Powell, CJCS.

We stood on deep pile carpet. I couldn't believe how soft it was or how clean it appeared. Military paintings were tastefully hung on the walls. After shaking hands, we moved to a corner where there were armchairs for all. The general's aide-de-camp offered us water. The major poured the ice-cold water from a pitcher into our glasses. We discussed the origins of the National Society of Scabbard and Blade. He was a gracious and intellectually curious general officer. Then he asked if we were familiar with the Pershing Rifles military fraternity.

I immediately spoke up and said, "Yes sir, I was a member of Company L 4th Regiment." "Where is that," he asked. I said, "North Carolina State University." He grinned and then said, "A-8, City College of New York." The ice, at least for me, had been broken at that moment. We had more in common than just our Ranger tabs.

Our allotted thirty minutes with the chairman was close to an end. It was time to execute the purpose for our visit. With Chairman Powell standing at attention, I draped the medal and neckband over his head to hang from his neck. With the draping, another in our party read aloud the award citation. The secretary shot photographs. We shook hands again as the general's aide-de-camp stood by holding three identical boxes. In appreciation for having bestowed the National Society of Scabbard and Blade's highest award upon him, the chairman presented each of us with a gift.

The cufflink—tie tac set was adorned with a miniature CJCS flag along with a special handwritten note on general officer stationery. This made my day. To my great surprise, several weeks later I received in the mail a photograph of me draping the award around the recipient's neck. General Powell had written on the photograph. "Thank you and best wishes Major Harris! Colin L. Powell"

SERENDIPITY

Upon arriving at Fort Bragg, North Carolina in early August 1990 for another training exercise, I learned that the Iraqi Army had invaded Kuwait the night before. The 82nd Airborne Division was in the midst of deploying an airborne infantry brigade. The advance party, including a good portion of the G2 from that division, was already in the air headed to Saudi Arabia. Those G2 members left in the rear were working nonstop to manage the collection of information flowing in from all sources of intelligence collection agencies. It was a busy shop.

The 29th's G2 training mission was immediately changed. Intelligence and imagery analysts began assisting an already grossly overloaded 82nd G2. It was a training opportunity that overnight turned into a real-world operation. We just happened to be in the right place at the right time.

The members of the 29th's G2 worked off and on for several weeks feeding information to the 82nd's G2 now in Saudi Arabia. The airborne infantry brigade was preparing a defense if the Iraqis attempted to assault them. In the meantime, another brigade of airborne infantry was being deployed. Soon, most of the deployable paratroopers from the 82nd would be in country. While this was in motion, the 29th's G2 continued to transmit reports of enemy movement in Kuwait — its service was invaluable to the boots on the ground.

History now records that an Iraqi invasion of Saudi Arabia might have been imminent, but Iraq's mobile forces never attacked, thus providing the time necessary for coalition forces to mass, train, and then push the Iraqis from Kuwait.

It became obvious to me that the 29th Infantry Division (Light) would not mobilize and deploy to the Persian Gulf for Operation Desert Shield when the state headquarters told us to alert our eight-gun battery of M198s that they were to be hand receipted for overseas deployment. The notification was received over the STU-III, a secure telephone, in my office.

Before Desert Storm ended, other weapon systems and air platforms were taken away. The division's lethality was being siphoned off. Then, suddenly out the blue, I was reassigned yet again.

PROMOTABLE

The Infantry Training Detachment (ITD) had always been a controversial unit from the beginning of the light division's rebirth. It was stationed at Fort A. P. Hill in central Virginia, a post of considerable acreage in a remote location of the Commonwealth.

Perhaps the main reason for all its shrouded mystery was the lack of understanding its mission, training sustainment and teaching specialty skills to members of a light division.

The division's G3 had make a strong case for the addition of a training detachment and for a new light infantry training center. The cadre were seasoned small unit leaders. All were qualified light leaders from the course taught by the infantry center at Fort Benning, Georgia. These highly skilled soldiers would become the division trainers for numerous schools including a Light Leaders Course, the Rites of Passage and a Light Fighters Course.

It was their purpose to mold all twenty-two battalions and special troops into the spirit of a light infantry division. In time these instructors would influence every soldier of the division. Many were inspired to attend the prestigious United States Army Ranger School.

The original commander, a ranger qualified veteran of the Vietnam War era, was challenged to recruit experienced soldiers for the purpose of becoming instructors. He hired very qualified non-commissioned officers. This instructor group schooled all the courses to sustain the skill sets of the light infantry division. The men were tough mentally and rough physically. They trained alongside of the Marine Corps' sniper instructors. Many were graduates of the sniper school at Quantico, Virginia and often assisted the east coast SEAL team in its inland mission which used the training areas for airborne insertions.

The instructor group was broken into teams that developed training lessons based on their experiences and assistance from Fort Benning, Georgia. These men were leaders at the squad and platoon levels and understood and expected all the respect due them by their peers. They had no contemporaries other than the SEALs that visited often enough from Little Creek, Virginia.

The ITD teams worked hard and played hard. They always attempted to out best each other whether on the job or off duty. The group required strong leadership from its commander and staff officers.

Once again, I was summoned by the chief. (He briefed me behind closed doors in his office on the IDT leadership situation.)

"Major Rock (pseudonym) has been relieved for cause. The long and the short of it is he directed that excess ammunition, rather than be returned to the ammunition bunker for future use, be hidden out of sight in a nearby lake. The post commander had the lake drained and discovered the ruined contraband. He was relieved over a waste, fraud, and abuse claim by one of his own staff officers. An investigation has already been conducted." The chief explained.

"What does this have to do with me?" I replied.

"The CG wants you to take command of the detachment. How fast can you get down there?" He asked.

The change of stations was the least of my concerns. Fort A. P. Hill was approximately and hour and half from Fort Belvoir where I maintained a residence. I was worried about the morale at the IDT, and I was also worried about the installation's leadership, specifically the attitudes of the post commander and the civilian executive officer.

As for me, I had worked more positions at division headquarters than I thought possible. Nearly six years had passed since I reported in at Building #2247, the original and provisional division headquarters. The G2 had been a fantastic experience, and I had been closely watching the development of the 629th Military Intelligence Battalion (CEWI). All that now changed as I headed back into an infantry assignment. The new position was commanded by a lieutenant colonel; I was promotable!

Epilogue

Within two years after taking command of the Infantry Training Detachment at Fort A. P. Hill, Virginia, the author was promoted to lieutenant colonel (LTC). Joe Harris remained here for about four years living in nearby Bowling Green, Virginia. He would eventually be reassigned to the headquarters of the 2nd Brigade, 29th Infantry Division (Light) also at Fort A. P. Hill. As the brigade executive officer, in 1995 he and the staff performed admirably at a division level Warfighter Exercise at Fort Leavenworth, Kansas. The brigade was deactivated in 1996, his final infantry assignment.

The author never returned to serve on the general staff of the 29th Infantry Division (Light). He did work on a corps level plan that was designed to exercise the division's command and control measures.

Joe ended his military service in the same style as it had begun as an ROTC cadet many years before, in a WWII barracks. Here, he removed his fatigues and boots for the final time. He retired from active service on September 1, 1998.

Endnotes

(1) Chapter 1, #268 episode: The Division Council consisted of four voting members; the Adjutants General of Virginia and Maryland, the Deputy Adjutant General of Virginia, and the Assistant Adjutant General for Army from Maryland were the initial councilmen. When the Division Commander was selected, the commanding general became the fifth voting member of the council. The Adjutant General of Virginia was the chairman during even numbered years, and the Adjutant General of Maryland was the chairman during odd numbered years. The division's chief of staff was appointed as the Secretary to the Division Council without vote.

(2) Chapter 2, Intent episode: James A. "Jim" Baber, III served as the initial division commander beginning in February of 1985. Jim was commissioned into the Field Artillery branch in 1952. He served on active duty for two years before transferring into the Ready Reserve. He joined the Army National Guard in 1954 and served in various command and staff assignments in the Field Artillery branch. In 1977 he transferred into the Infantry branch. He served as deputy brigade commander and brigade commander of the 116th Infantry Brigade (Separate). He was the Deputy Adjutant General at the time of his selection as Commanding General, 29th Infantry Division (Light). Major General Baber was an attorney at law in civilian life.

(3) Chapter 3, Charter episode: Bruce Jacobs joined the army during WWII and served in the Pacific Theater. He

spent the next four decades in active and reserve duty and service in the Army National Guard. Many assignments focused on public affairs work. He was a Korean War veteran and went to Vietnam in the late '60s as a special National Guard Liaison Officer. In retirement, he spent ten years with the National Guard Association and was editor of its magazine. He also served as executive director of the Historical Society of the Militia and National Guard, now called the National Guard Educational Foundation. He authored several books on military history.

(4) Chapter 3, Sojourner episode: After the North Dakota Regiment departed the Philippines, a group of Masons organized an informal club named "Sojourners." In 1901, these Sojourners provided the impetus, funds and paraphernalia to charter Manila Lodge under the Grand Lodge of California. The Philippine "Sojourners Club" resurfaced in 1907, open to Masons who were not members of local lodges; within a year it had nearly 200 members from every state and territory and eight foreign countries. Most of the Philippine Sojourners eventually became scattered throughout the US. A group of these military Masons sojourning in Chicago in 1917 formally organized the "Chicago Sojourners Club." Additional clubs were quickly formed at army posts and naval bases under the name "National Sojourners Club." In 1927, the word "Club" was dropped and it became "National Sojourners, Incorporated' in 1931.

(5) Chapter 4, Busted episode: On April 4, 1975 a Lockheed C-5A Galaxy crashed while returning to Tan Son Nhut Air Base on the outskirts of Ho Chi Minh City (formerly Saigon, South Vietnam). This flight of Operation Babylift was bound for Clark Air Base in the Philippine Islands, however, upon takeoff, a serious maintenance

malfunction forced the pilot to return to Tan Son Nhut. It crash-landed in a rice paddy killing 138 of 314 souls on board.

(6) Chapter 4, Cufflinks episode: Colin Powell joined the army in 1958 upon receiving his commission as a 2nd lieutenant through the Department of Military Science (Army ROTC) at the City College of New York. While an undergraduate, he pledged and was initiated into the National Society of Pershing Rifles. As an ROTC cadet he rose to the rank of colonel. He is a Distinguished Military Graduate. His storied army career of leadership positions led him to ultimately work as a White House Fellow in 1972. He was selected to be the 16th National Security Advisor (1987-1989) during the Reagan Administration and the 12th Chairman of the Joint Chiefs of Staff (1989-1993) by President George Herbert Walker Bush until his retirement from the army in 1993. In 2001 he was appointed the 65th Secretary of State by President George Walker Bush.

Appendix

The 29th Infantry Division (Light) Activation, Restructuring and Organizational Living Concept Plan

Blue and Gray Division 22 May 1985

EXECUTIVE SUMMARY

1. The Division's G6, Transition Office, acts as the umbrella for the entire conversion process. It's responsibilities include the monitoring and the coordination of all activities associated with the development of the 29th Infantry Division (Light). Of particular interest to the G6 is the monitoring of the carrier unit identification codes. To date, a CUIC is in effect at HHC 29 ID (L) with July 1985 set as the date for all other new units to receive their CUIC. A consolidated (overall) flow chart which outlines the sequence of activities for the transition of the Virginia and Maryland Army National Guard is found in Annex A (Transition Planning).

2. The restructure of the Virginia and Maryland Army National Guards with the Light Infantry Division concept will commence with all efforts directed to reducing personnel turmoil (reassignments and unit locations), avoiding excess equipment transfers and

reducing the need for new and/or expanded facilities. To this end, both states have submitted their draft stationing plans and the G6 has produced a consolidated 29th ID (L) Stationing Plan. It is anticipated that only minor changes will have to be made to the Division Stationing Plan.

3. A comparison of current authorized and assigned personnel in existing units against the Light Infantry Division design was conducted upon the receipt of the LTOE. The analysis identified the number of troops and appropriate grade and job skills in excess of and short of meeting the light infantry LTOE. The MTOE is scheduled to be distributed in the June/July time period and a final review will be conducted at that time. Priority for personnel fill within the Division will be directed at the newly established units with emphasis placed on the combat arms.

4. Identification of the MOS Q status of the Division is dependent upon receiving the MTOE's. Both the G1 and G3 are working to develop the current MOS Q of units forming the 29th ID (L) and a projection of a total or overall, MOS Q for the entire Division. Input to this projection will eventually require unit level information which will allow us to develop MOS training requirements. In order to define the Division's Yearly Training Program (YTP), a training strategy had to be developed and approved. The Division's training objectives for TY 86-87 are categorized below:

   a. Individual Training

      (1) Light Leader Course (LLC) for the cadre, E5-03 of infantry units, artillery FIST

team chiefs, and engineer unit cadre, in that order. The LLC is a Divisional School. Allocations are provided by the G3.

(2) Divisional Rites of Passage (ROP) for all assigned personnel. This will be for individuals or units. The ROP training is a Divisional School. The POI and method to accomplish this training has been provided to all the brigade S3s and commanders. Training dates, locations and allocations are provided by the G3.

(3) Primary Leadership Develop Course (PLDC) training will adhere to the respective Adjutant General's guidance.

(4) Ranger Training is a priority. The projected requirements for Ranger School quotas for FY 86 are due to this headquarters NLT 1 June 1985. The Division will start a two-week pre-Ranger Course during FY/TY 87. Allocations for personnel to attend the pre-Ranger Course will be provided by the G3.

(5) Physical Readiness Training. 30 minutes of formal PT will be conducted each training day. This requirement may be waived by the commander when in the tactical field environment. Ensure that only persons with a valid medical reason are excused. Use FM 21-20 to develop your unit's PT program.

(6) Weapons Qualifications. The Division commander's minimum standard for individual weapons qualification is sharpshooter. Units will normally use LLC qualified or marksmanship team personnel to conduct qualification on KD ranges during the training year. Units may use up to a MUTA 5 for weapons firing and schedule additional firing as required, provided ammunition is available. Personnel attending the Division LLC program will complete weapons qualification during the LLC training.

b. Unit Training

(1) Activating new units will accomplish the following training requirements at a minimum:

(a) Unit established on E-DATE

(b) Conduct IDT, within 30 days of carrier E-DATE, and publish IDT schedule for remainder of TY.

(c) Program personnel to attend LLC, ROP, and LFC as required.

(d) Program selected personnel to attend PLDC as required.

(e) Identify individual MOS training requirements.

(f) Plan for and attend AT 86 (26 July-9 August), publish 2 year training calendar NLT 30 September.

(g) Conduct section training.

(h) Conduct training for individual physical readiness and maintenance of current MOS qualification.

(i) Attain MOS qualification IAW NGR 10-1.

(j) Submit USR as required.

(k) Prepare for and pass unit's Federal Recognition Inspection.

(2) Existing and reorganized units will use the following Division assumptions and planning objectives to develop their training plans.

(a) Assumptions:

1 Funding ammunition, training areas and 55 space training TDA for this Division are available in TY 86.

2 Division budget, submitted through STARC-VA to NGB is approved.

(b) Planning Objectives:

1 Identify TY 87 unit schools NLT May 86; e.g., LLC, ROP, etc.

2 Publish TY 86-88 training guidelines NLT 30 June 1985.

3 Publish TY 87-89 training guidelines NLT 30 June 1985.

4 AT 85 for 58th SIB and 116th SIB (13-27 July 1985).

5 Identify school requirements for TY 87 for USAR Schools, Army Service Schools, and Unit Schools NLT 30 September 1985.

6 Complete LLC I (January 1985-29 September 1985).

7 Plan for AT 86 (Div (-) 29 July-9 August 1986) and (DIVARTY 14 -28 June 1986), AT 87 and 88 in 4th Quarter.

8 Conduct Division Rites of Passage for 2,000

personnel/year, TY 86-91 (IDT), 4 day POI.

9 Conduct Division LLC II for up to 300 personnel/year, TY 86-91 (IDT) and (AT), 28 day POI.

10 Infantry units conduct LFC TY 86-88 with Plt ARTEP concurrent, 21 day POI.

11 Engineer units conduct SAPPER Course TY 87-88.

12 Division achieve C-3 (65% of required strength) MOS qualification (Oct 89).

13 Conduct Co/Btry external ARTEPs (TY 89).

14 Initial Division Rites of Passage complete (TY 89).

15 LFC training complete (Infantry) (TY 88).

16 Engineer SAPPER training complete (TY 88).

17 NET concurrent to ongoing programs based on new equipment (TY 86-91).

(c) All Division units will be prepared to participate in or support the 29th Infantry Division (Light) Reactivation Ceremony on 4-5 October 1985 IAW OPORD 3-85.

(d) Infantry units with LLC qualified Cadres will execute Rites of Passage (ROP) and Light Fighters Course (LFC) training.

5. The overall flow chart depicted at Annex A (Transition Planning) includes a diagramming of the new installation which has been prepared to house the 29th ID (L) Headquarters. The request for the proposed new 29th ID (L) Training Center was drafted in November and subsequently forwarded to NGB in December of last year. Enclosures located in Appendix 2 of Annex A focus on the identification and occupation of temporary facilities at various locations. Early identification of the equipment requirements is necessary to provide for ease and efficiency upon restructuring the units in the field. The G4 has identified a schedule by which the equipment fielding plan must be executed.

6. The Activations, Restructuring and Organizational Living Concept Plan provides a detailed analysis of the events and activities that must be accomplished by each staff section in order to convert the Virginia and Maryland Army National Guards into the 29th Infantry Division (Light). The Wartime Mission of each SIB remains unchanged through FY 86. Restructure actions involve reorganization (conversion) and activation of units during the

transition period. Struct adherence to the scheduling of events on every flow chart is the keystone to a smooth transition to a light infantry division and to assuming a new Wartime Mission in FY 87. The hiring of the full-time staff (FTM-AGR) is on track. The Division Headquarters is staffing up along with the brigades and battalions. Brigade and battalion commanders or designees have been earmarked for the new division. Selection of commanders at the company, troop and battery level is ongoing. Throughout the transition, each state will continue to give as a top priority the training necessary to maintain unit readiness if a mobilization occurs under the current Wartime Mission. Concurrently, the 29th Infantry Division (Light) is striving to achieve the status of becoming functional, elite and totally combat ready prior to assuming its Wartime Mission.

7. The POC for this headquarters is Major Joe E. Harris, Jr., Autovon 354-4732.

Acknowledgements

I knew no one except for a few individuals in Richmond, Virginia and the one who hired me sight unseen over the telephone. Walking into the provisional headquarters on that cold January day, I was all eyes and ears. Sergeant Will Mund was the first soldier I met. It was one of those unforgettable moments. In retirement Will has been a loyal and dedicated servant to the 29th Division Association, and to the Maryland Military Historical Society.

Retired Brigadier General William C. Bilo was given a monumental assignment, develop a G-staff capable of managing a light infantry division. He listened to a recommendation and took a chance on me. I thoroughly enjoyed division level challenges. It was a tremendous learning experience. I appreciate the opportunity to serve in this great endeavor. Thank you chief!

My good friend and retired Colonel Jay W. Lytle is another individual I wish to recognize. Jay, you provided sage advice to me on a number of personal and professional matters. Thank you for your interest in my welfare, and many thanks for your straight forward professional guidance. Because of you, my introduction to the world of intelligence was a very rewarding experience.

Many of the folks I encountered during the five years I spent at Fort Belvoir have passed on to their reward. All of them touched my life someway. I owe a debt to each of them for their efforts to make this light division a force for its members to proudly say, "Twenty-Nine Let's Go!"

Working and residing on an army post is a unique experience. I enjoyed the years I lived in government quarters on Fort Belvoir. I embraced the culture, customs, and atmosphere that are very different from the civilian world. Living as well as working on a post had advantages and its disadvantages, but in the end my neighbors—the army family—provided me a richer appreciation of military life whether in the field or the garrison.

I also have to thank the team at JWC Publishing. Tim Jacobs, Michael Salter, and Brian Mast answered my every question and I'm so pleased with the final result. Thank you, guys!

About the Author

Lieutenant Colonel Joe E. Harris, Jr. entered active duty in June 1972 after graduating from North Carolina State University. He was a Distinguished Military Student and is a Distinguished Military Graduate. He was commissioned into the Regular Army.

Branched Infantry, Lieutenant Colonel Harris commanded platoons, a mechanized rifle company, and a division level light Infantry Training Detachment. He served in many staff positions and was also trained in the Military Intelligence Branch. He has served in the Regular Army, the United States Army Reserve, the North Carolina Army National Guard and the Virginia Army National Guard.

His awards and decorations include numerous state awards, an Army Commendation Medal with two Oak Leaf Clusters, a Meritorious Service Medal with one Oak Leaf Cluster and the Legion of Merit. His skill badges include the Ranger Tab, Expert Infantryman Badge, Parachute and Air Assault Badges.

Born 1949 in Elkin, North Carolina, storytelling seemed natural having listened to many raconteurs from the foothills of the Blue Ridge Mountains. His experiences in the reactivated 29th Infantry Division (Light) headquarters convey stories of humor, frustration, joy and sadness. In *No Two Days the Same: Bringing Back a Division*, he reminisces about his mid-career journey on the G-Staff while residing on Fort Belvoir, Virginia.

Adopting Arizona as his state of residence, Joe currently lives outside of Tucson in Oro Valley.

Made in the USA
Monee, IL
04 November 2026

67380775R00069